Contents

Foreword

The U.S. Environmental Protection Agency (EPA) is charged by Congress with protecting the Nation's land, air, and water resources. Under a mandate of national environmental laws, the Agency strives to formulate and implement actions leading to a compatible balance between human activities and the ability of natural systems to support and nurture life. To meet this mandate, EPA's research program is providing data and technical support for solving environmental problems today and building a science knowledge base necessary to manage our ecological resources wisely, understand how pollutants affect our health, and prevent or reduce environmental risks in the future.

The National Risk Management Research Laboratory (NRMRL) is the Agency's center for investigation of technological and management approaches for preventing and reducing risks from pollution that threaten human health and the environment. The focus of the Laboratory's research program is on methods and their cost-effectiveness for prevention and control of pollution to air, land, water, and subsurface resources; protection of water quality in public water systems; remediation of contaminated sites, sediments and ground water; prevention and control of indoor air pollution; and restoration of ecosystems. NRMRL collaborates with both public and private sector partners to foster technologies that reduce the cost of compliance and to anticipate emerging problems. NRMRL's research provides solutions to environmental problems by: developing and promoting technologies that protect and improve the environment; advancing scientific and engineering information to support regulatory and policy decisions; and providing the technical support and information transfer to ensure implementation of environmental regulations and strategies at the national, state, and community levels.

This publication has been produced as part of the Laboratory's strategic long-term research plan. It is published and made available by EPA's Office of Research and Development to assist the user community and to link researchers with their clients. It should be of particular assistance to those planning and/or designing an urban watershed management campaign.

Sally G. Gutierrez, Acting Director
National Risk Management Research Laboratory

Preface

Stormwater is part of a natural hydrologic process. However, human activities, especially those in an urban environment, cause significant changes in patterns of stormwater flow from land into receiving waters. The undesirable impacts of stormwater runoff can be controlled by prudent management efforts, referred to as best management practices (BMPs). The purpose of this report is to provide a general information on the most commonly used structural BMP options, the design considerations involved, and the general guidelines for monitoring, selection, implementation, and associated costs of BMPs in urban watersheds.

Urban land area in the U.S. more than quadrupled between 1945 and 1997, increasing from 15 to 64 million acres, primarily from conversion of land that was previously forested, pastured, and ranged, while during this same period, paved road mileage nearly tripled. This urbanization, with the accompanying expansion of impervious surfaces, significantly increased the nation's total amount of developed and paved surface area; reduced the amount of wetlands and other undeveloped land; and, decreased our watersheds' ability to mitigate floods, facilitate sediment replenishment, and protect water quality and aquatic health by removing excess nutrients and other chemical contaminants before runoff enters receiving waters. Urban development also causes changes in the hydrologic regime, which can result in habitat loss and other ecological impairments due to more frequent and longer bankful events, lower baseflows, and increased stream channel erosion. Significant decreases in biotic measures of aquatic ecologic health also become apparent at low relative impervious areas.

Urban areas are often proximal to highly-valued recreational and environmentally-significant waters, critical estuarine and coastal areas, as well as lakes and reservoirs that supply many drinking water systems. The water quality impacts associated with this increasing urbanization are evident in the impairments facing many of these waters. The U.S. EPA's biennial report, The National Water Quality Inventory, published in 2002, indicates that the states collectively believe that about 40% of assessed U.S. streams, lakes, and estuaries fail to meet one or more designated uses. As in previous inventories, pollution from urban land, transported by precipitation and runoff, is identified as a leading source of impairment, and many of the common stormwater-associated stressors are identified as the leading causes of impairment, including: siltation, nutrients, bacteria, metals, and oxygen-depleting substances.

During the past thirty years, the U.S. EPA established several regulatory programs to address the various point and diffuse sources of urban runoff. After passage of the Clean Water Act (CWA) in 1972, water pollution control efforts focused primarily on wastewater discharges from point sources consisting of municipal treatment plants, industrial plants, and combined sewer overflows (CSO). To bring attention to the latter, the U.S. EPA issued a National CSO Control Strategy on August 10, 1989 reaffirming CSOs as point-source discharges subject to CWA National Pollution Discharge Elimination System (NPDES) permit requirements (Federal Register, 1989). Although the Strategy successfully focused attention on CSOs, it fell short in resolving many fundamental impediments to its implementation. In mid-1991, the U.S. EPA began a process to accelerate CSO strategy implementation. The initiative resulted in a CSO Control Policy which, in addition to providing guidance to NPDES permitting and enforcement authorities, also provided municipalities, especially those that are financially disadvantaged, the flexibility to consider the site-specific nature of CSOs and determine the most cost-effective means of reducing pollutants to meet the requirements of the CWA.

Far less emphasis had been placed on diffuse source pollution, i.e., pollution from contaminants picked up and carried into surface waters by stormwater runoff. To address the numerous environmental stressors associated with this pollution source, the CWA was amended in 1987 to establish NPDES stormwater discharge requirements. To implement these requirements, the U.S. EPA published the "Phase I" Stormwater permit program to address certain stormwater discharge categories associated with industrial activities and large storm sewer systems (located in 842 municipalities with populations of 100,000 or more) on November 16, 1990. On December 8, 1999, the U.S. EPA promulgated "Phase II" stormwater regulations expanding the list of regulated communities to include small systems in Bureau of Census-defined urbanized areas. Phase II requires NPDES permits and stormwater discharge controls in the regulated communities.

The U.S. EPA's approach, as outlined in the Phase II regulations, is to control the stormwater stressor load using best management practices (BMPs) augmented by a combination of natural and assisted restoration within the receiving water. The regulations rely heavily on a combination of structural and nonstructural techniques (i.e., BMPs) to mitigate the stormwater effects on receiving waters. Structural BMPs are designed to trap and detain runoff to settle or filter out the constituents before they enter receiving waters, while nonstructural BMPs control pollutants at the source to prevent or reduce contamination of stormwater runoff. There are many types of structural BMPs available for stormwater runoff control and storage-treatment, and new treatment devices are frequently being introduced; however, there is a limited understanding of their effectiveness, their lifetime cost, the appropriate techniques to design or position them in the watershed, or their impact on receiving water after installation. The understanding of nonstructural BMPs is even more limited.

The U.S. EPA's Urban Watershed Management Research Program (UWMRP) believes that efforts accumulating additional data measuring BMP-specific performance as relative stressor reductions will add little to the knowledge of BMPs, the internal physical and chemical mechanisms that control stressor and multi-stressor removals, or the ability to transfer knowledge to other BMP sites. Alternatively, establishing techniques that enable communities to predict the effluent quality *can* be translated to any urban installation. Providing tools and techniques that allow community-based environmental planners to select cost-effective techniques and practices that can restore and protect the receiving water body from stormwater-associated stressors on a watershed scale and within a predictable time is the UWMRP's overall objective for water quality research.

The UWMRP is emphasizing cost-effective technologies, and structural and nonstructural BMPs that reduce the impact of wet-weather flow related pollutants from *urban* areas. In the urban environment, wet-weather pollution can emanate from CSOs, sanitary sewer overflows (SSOs), and stormwater runoff. The UWMRP has a long history of involvement and expertise in CSO and SSO control as well as in stormwater runoff control. Early work in control technologies focused on developing and demonstrating: advanced collection system design alternatives; high-rate treatment technologies; techniques to optimize flow capture and treatment through in-sewer storage, flow routing, and wastewater treatment plant operation; and, infiltration/inflow identification and elimination technologies.

The removal of pollutants from WWF using control technologies can be significantly expensive. BMPs provide a mostly passive, and less costly, means to mitigate the effects of stormwater on receiving waters and have become a ubiquitous part of the landscape. Historically, engineers installed these controls for runoff volume (flood) control, and have only recently extended the application to include water quality considerations. During the comparatively short period that BMPs have been applied to water quality, researchers have collected a substantial amount of data in an attempt to evaluate their performance. In addition, government agencies and professional organizations have published numerous guidance documents to assist in their implementation. However, the menu of structural and nonstructural BMP options is extensive and there is little consistency in the amount or quality of design, performance, and cost data for each of the many practices currently in use.

The purpose of this document is to provide a general description and insight on the various BMP options, the design considerations involved, and the general guidelines for selection, implementation, monitoring and associated costs of BMPs in urban watersheds. The main focus of the document is on structural BMPs; the various nonstructural practices are discussed only briefly. It is a comprehensive literature review on commonly used urban watershed BMPs that heretofore was not consolidated. The document does not intend to dictate or specify the actual selection of BMPs but attempts to provide the framework for an informed selection of BMPs for any stormwater management program. It was prepared to serve as an information source to individuals and agencies, municipalities, watershed management groups, etc. on the existing state of BMPs in urban stormwater management.

This document has been reviewed in accordance with the U.S. Environmental Protection Agency's peer and administrative review process, and has been approved for publication as a U.S. EPA document. Any mention of products or trade names in this book does not constitute recommendation for use by the U.S. EPA. Any opinions expressed are those of the author(s) and do not, necessarily, reflect the official positions and policies of the U.S. EPA. Any information within the document should not be considered as official U.S. EPA guidance. In addition, all reported recommendations or values do not reflect the views of the agency and are solely owned by the cited sources.

Anthony N. Tafuri, P.E.
Chief, Urban Watershed Management Branch
Water Supply & Water Resources Division
National Risk Management Research Laboratory

Acknowledgments

A book of this nature results from the experiences of many and, in particular, from the volunteer efforts of a dedicated few to whom we are grateful:

The U.S. Environmental Protection Agency (U.S. EPA), Water Supply and Water Resources Division (WSWRD), National Risk Management Research Laboratory (NRMRL), Office of Research and Development (ORD), which provided for the extremely important Wet-Weather Flow Management Research Program without which this book would not have been written.

Our colleagues, who, over the last four decades, have given us the opportunity to learn from them.

Our peer reviewers: Eric Strecker, PE, Geosyntec Consultants; Paul Zambratto, and Maureen Krudner, NYC Watershed Team, U.S. EPA Region 2; and, Jonathan D. Essoka, Ph.D., U.S. EPA Region 3, for the critical review of this book. We thank Ronald B. Landy, Ph.D., U.S. EPA Region 3, for his help in coordinating the reviews from Region 2 and 3 for this book. Our special thanks to Michael Borst, Asim Ray, Ph.D., and Joyce Perdek Walling, PE, Urban Watershed Management Branch (UWMB), and Marie O'Shea, Ph.D., U.S. EPA Region 2, for reviewing this book and providing valuable comments.

Richard Koustas, U.S. EPA, Region 2, and Ramjee Raghavan, Ph.D., PE, USInfrastructure, Inc., for their input to Chapter 6 - BMP Costs.

Judy Norinsky and Jatu Bracewell, Interns, Environmental Careers Organization, for their assistance in editing.

Jaime Marin, Computer Sciences Corporation, Inc., who provided computer technical support to convert chapters to the publisher's format.

List of Select Acronyms

ASCE	American Society of Civil Engineers
BMP	best management practice
BOD	biochemical oxygen demand
Caltrans	California Department of Transportation
Cd	cadmium
COD	chemical oxygen demand
Cp_v	channel protection storage volume
CSO	combined sewer overflow
Cu	copper
CZARA	Coastal Zone Act Reauthorization Amendments
CZMA	Coastal Zone Management Act
DO	dissolved oxygen
ED	extended-detention
EMAP	Environmental Monitoring and Assessment Program
EMC	event mean concentration
ENR	Engineering News Record
FC	fecal coliform
FDEP	Florida Department of Environmental Protection
FHWA	Federal Highway Administration
ISMDSF	integrated stormwater decision-support framework
LID	low-impact development
MCTT	multi-chambered treatment train
MEP	maximum extent practicable
MS4	municipal separate storm sewer systems
NGPE	natural growth protection easement
NH_3	ammonia
NJDA	New Jersey Department of Agriculture
NO_2^-	nitrite
NO_3^-	nitrate
NPDES	National Pollutant Discharge Elimination System
ASCE	Natural Resources Conservation Service
NURP	Nationwide Urban Runoff Program
NVPDC	Northern Virginia Planning District Commission
NYCDEP	New York City Department of Environmental Protection
NYSDEC	New York State Department of Environmental Conservation
O&M	operation and maintenance
PAHs	polycyclic aromatic hydrocarbons
Pb	lead
PIT	pilot infiltration test
QA/QC	quality assurance/quality control

QAPP	quality assurance project plan
Re_v	recharge volume
Rv	runoff coefficient
SS	suspended solids
SSC	suspended solids concentration
SWPPP	stormwater pollution prevention plan
TKN	total kjeldahl nitrogen
TMDL	total maximum daily load
TP	total phosphorus
TSS	total suspended solids
U.S. EPA	United States Environmental Protection Agency
USDA	United States Department of Agriculture
USGS	United States Geological Survey
WEF	Water Environment Federation
WQ_v	water quality volume
WWF	wet-weather flow
Zn	zinc

1 The Use of Best Management Practices (BMPs) in Urban Watersheds - Executive Summary

Swarna Muthukrishnan, Bethany Madge, and Ari Selvakumar

CONTENTS

1.1 INTRODUCTION

Diffuse source pollution is one of the most intricate environmental problems with extensive impacts on surface and groundwater quality. It is a major factor impacting the quality of water supply, and the rate at which diffuse source pollutants are generated and delivered to water resources is greatly affected by anthropogenic activities as well as natural processes. The main hydrologic component transporting these pollutants to surface water bodies is runoff, which results from precipitation or snowmelt (Leeds *et al.*, 1993). Stormwater is part of a natural hydrologic process; however, human activities, especially urban development and agriculture, cause significant changes in patterns of stormwater flow from land into receiving waters. Urban runoff can be or is often a significant source of water pollution, causing decline in fisheries, swimming, and other beneficial attributes of water resources (U.S. EPA, 1993). Urban stormwater runoff includes all flows discharged from urban land uses into stormwater conveyance systems and receiving waters; in this context, urban runoff includes both dry- weather non-stormwater sources (e.g., runoff from landscape irrigation, dewatering, and water line and hydrant flushing) and wet-weather stormwater runoff. Water quality can also be affected when runoff carries sediment and other pollutants into streams, wetlands, lakes, estuarine and marine waters, or groundwater. The costs and impacts of water pollution from urban runoff are significant and can include:

- fish kills,
- health concerns of human and/or terrestrial animals,
- degraded drinking water,
- diminished water-based recreation and tourism opportunities,

- economic losses to commercial fishing and aquaculture industries,
- lowered real estate values,
- damage to habitat of fish and other aquatic organisms,
- inevitable costs of clean-up and pollution reduction,
- reduced aesthetic values of lakes, streams, and coastal areas, and
- other impacts (Leeds *et al.*, 1993).

Increased stormwater flows from urbanization have the following major impacts (FLOW, 2003):
- acceleration of stream velocities and degradation of stream channels,
- declining water quality due to washing off of accumulated pollutants from impervious surfaces to local waterways, and an increase in siltation and erosion of soils from pervious areas subject to increased runoff,
- increase in volume of runoff with higher pollutant concentrations that reduces receiving water dilution effects,
- diminished groundwater recharge, resulting in decreased dry-weather flows; poorer water quality of streams during low flows; increased stream temperatures; and, greater annual pollutant load delivery,
- increased flooding,
- combined and sanitary sewer overflows due to stormwater infiltration and inflow,
- damage to stream and aquatic life resulting from suspended solids accumulation, and
- increased health risks to humans from trash and debris which can also endanger and destroy food sources or habitats of aquatic life (FLOW, 2003).

The major categories of stormwater pollutants, their sources, and related impacts are presented in Table 1-1.

Table 1-1. Major Categories of Stormwater Pollutants, Sources and Related Impacts

Stormwater Pollutant	Major Sources	Related Impacts
Nutrients: Nitrogen, Phosphorus	Urban runoff; failing septic systems; croplands; nurseries; orchards; livestock operations; gardens; lawns; forests; fertilizers; construction soil losses	Algal growth; reduced clarity; lower dissolved oxygen; release of other pollutants; visual impairment; recreational impacts; water supply impairment

Solids: Sediment (clean and contaminated)	Construction sites; other disturbed and/or non-vegetated lands; road sanding; urban runoff; mining operations; logging operations; streambank and shoreline erosion	Increased turbidity; reduced clarity; lower dissolved oxygen; deposition of sediments; smothering of aquatic habitat including spawning sites; sediment and benthic toxicity
Oxygen-depleting substances	Biodegradable organic material such as plant; fish; animal matter; leaves; lawn clippings; sewage; manure; shellfish processing waste; milk solids; other food processing wastes; antifreeze/other de-icing chemicals; other applied chemicals	Suffocation or stress of adult fish, resulting in fish kills; reduction in fish reproduction by suffocation/stress of sensitive eggs and larvae; aquatic larvae kills; increased anaerobic bacterial activity resulting in noxious gases or foul odors often associated with polluted water bodies; release of particulate bound pollutants
Pathogens: Bacteria, Viruses, Protozoans	Domestic and natural animal wastes; urban runoff; failing septic systems; landfills; illegal cross-connections to sanitary sewers; natural generation	Human health risks via drinking water supplies; contaminated shellfish growing areas and swimming beaches; incidental ingestion or contact
Metals: Lead, Copper, Cadmium, Zinc, Mercury, Chromium, Aluminum, others	Industrial processes; mining operations; normal wear of automobile brake pads and tires; automobile emissions; automobile fluid leaks; metal roofs; gutters; landfills; corrosion; urban runoff; soil erosion; atmospheric deposition; contaminated soils	Toxicity of water column and sediment; bioaccumulation in aquatic species and through food chain

Hydrocarbons: Oil and Grease, Polyaromatic hydrocarbons (PAHs) - e.g., Naphthalenes, Pyrenes	Industrial processes; automobile wear; automobile emissions; automobile fluid leaks; waste oil	Toxicity of water column and sediment; bioaccumulation in aquatic species and through food chain; lower dissolved oxygen (DO); coating of aquatic organism gills/impact on respiration
Organics: Pesticides, Polychlorinated biphenyls (PCBs), Synthetic chemicals	Applied pesticides (herbicides, insecticides, fungicides, rodenticides, etc.); industrial processes; nurseries; orchards; lawns; gardens; historically contaminated soils/wash-off	Toxicity of water column and sediment; bioaccumulation in aquatic species and through food chain
Inorganic Acids and Salts (sulphuric acid, sodium chloride)	Irrigated lands; mining operations; landfills; road salting and uncovered salt storage	Toxicity of water column and sediment

(Leeds *et al.*, 1993; MA DEP and MA CZM, 1997; U.S. EPA, 2000)

1.2 The Concept of BMPs

The undesirable impacts of stormwater runoff can be controlled by prudent management efforts. Stormwater management encompasses an array of measures that involve careful application of site design principles, construction techniques to prevent sediments and other pollutants from being released and/or entering surface or groundwater, source controls, and treatment of runoff to reduce pollutants and reducing the impact of altered hydrology.

For many years, federal and state regulations for stormwater management efforts were oriented towards flood control with minimum measures directed towards improving the quality of stormwater such as sediments and erosion control and the reduction of pollutants. The U.S. recognized the problem of diffuse pollution many years ago and established provisions in a major amendment to the Clean Water Act in 1987, leading to national programs of action to address the issue. The increased awareness of the need to improve water quality in the last two decades resulted in the concept of best management practices (BMPs) which are measures intended to provide an on-the-ground practical solution to diffuse pollution problems from all sources and sectors (D'Arcy and Frost, 2001). BMPs are technology and education based requirements in the federal stormwater regulations that call for the implementation of controls to reduce the discharge of pollutants to the Maximum Extent Practicable (MEP) in municipal-type stormwater systems (Caltrans, 2002). BMP refers to operational activities, physical controls or educational measures that are applied to reduce the discharge of pollutants and minimize potential impacts upon receiving waters, and accordingly, refers to both structural and nonstructural practices that have direct impacts on the release, transport, or discharge of pollutants.

The proper management of stormwater runoff is necessary to reduce stream channel erosion, pollution, siltation, sedimentation, and local flooding, all of which have adverse impacts on the land, water resources, and the people. The BMP program was increasingly designed in the 1980s primarily to address pollution from wet-weather flow (WWF) and polluted runoff and focused on controlling runoff increases and reducing water quality degradation associated with new development. The goal of these practices is to maintain the predevelopment characteristics as close as possible, even after development of a site, and/or to reduce the impacts to an accepted level. It must be understood in this context that BMPs do not merely act as controls for new development, but these practices equally apply to existing developments as well as areas that have undergone any kind of re-development.

The BMP concept has the following key elements (D'Arcy and Frost, 2001):

- There is a need for guidance that offers practical prevention options.
- The options need to be defined and explicit best practice rather than ill-defined individual interpretations of what is required.
- The options should be describable as best practice, based on research and experience.

Since the development of BMPs, various state and local governments have adopted a profusion of laws, regulations, and policies to encourage or mandate the use of urban BMPs. These BMPs have been developed and refined to mitigate some, if not all, of the adverse impacts associated with any kind of development/re-development activity. The capabilities of each BMP are unique. This needs to be recognized along with its limitations, and these factors, in addition to the physical constraints at the site, need to be judiciously balanced with the overall management objectives for the watershed in question. At a minimum, a BMP program developed for a site should strive to accomplish the following set of criteria:

- Reproduce, as nearly as possible, the natural hydrological conditions in the stream prior to development or any previous human alteration (Schueler, 1987; Young *et al.*, 1996).
- Provide a moderate to high level of removal for most urban pollutants as one of a set of BMPs in the watershed working together to achieve desired receiving-water quality.
- Be appropriate for the site, given physical constraints.
- Be reasonably cost-effective in comparison with other BMPs.
- Have an acceptable future maintenance burden.
- Have a neutral impact on the natural and human environment.

The purpose of this book is to provide a general description and insight on the various BMP options, the design considerations involved and the general guidelines for selection, implementation, and monitoring of BMPs to reduce pollutants in urban stormwater from new development and re-development. As the main focus of this book is structural BMPs, the various nonstructural practices is discussed only briefly. This book however, does not intend to dictate or specify the actual selection of BMPs, but attempts to provide the framework for an informed selection of BMPs for any stormwater management program.

1.3 SUMMARY AND FINDINGS OF THE BOOK

Chapter 2 provides a general discussion of the most commonly used nonstructural and structural BMPs for the management of urban storm runoff. The introduction defines what a BMP is and describes what structural and nonstructural BMPs are. The chapter discusses how the distinction between structural and nonstructural BMPs is quite clear in many cases, and not entirely so in the case of some other BMPs. Some good examples related to emerging concepts in runoff management are presented, such as better site design and Low-Impact Development (LID) techniques e.g., green roofs (Section 2.1), that focus on the use of both site planning and small- scale treatment approaches. These practices reinforce the growing opinion that a combination of both structural and nonstructural practices in a treatment train approach is almost certainly a better option to meet stormwater management objectives for many project sites. Section 2.2 briefly describes the types of structural BMPs currently being used and recommended by the U.S. EPA's menu of BMPs (U.S. EPA, 2001) and several state agencies. The general advantages and limitations of each of these BMPs are also discussed. Section 2.3 focuses on the major physicochemical and biological processes in BMPs and how they influence the removal of pollutants and mitigate other stressors from stormwater, and section 2.4 presents a brief overview of the factors influencing the performance of these structural BMPs.

In Chapter 3, Structural BMP Design Practices, the factors that need to be considered in designing urban BMPs are discussed. The following eight BMPs commonly used for stormwater treatment in new development are addressed: (i) dry extended-detention ponds; (ii) wet ponds; (iii) stormwater wetlands; (iv) grassed swales; (v) vegetated filter strips; (vi) infiltration trenches; (vii) porous pavement; and, (viii) sand and organic filters. As an introduction, Section 3.1 traces the development of the concept of BMPs from earlier stormwater management measures that first relied heavily on flood and then water quantity control to the current focus on controlling both the quality and quantity of runoff in order to mitigate the impacts to receiving waters. Section 3.2 briefly explains the sizing criteria involved in BMP design considerations and describes the performance objectives of these BMPs. Section 3.3 describes the design considerations for the above-mentioned eight BMPs in detail in separate subsections 3.3.1 - 3.3.8. Each subsection on a specific BMP carries a brief description of the BMP and how it addresses the two issues of stormwater control as well as pollutant removal. Each subsection also has a detailed presentation on the general design considerations including, site suitability, physical specifications, and geometry.

The design and construction of stormwater BMPs is a constantly evolving process in that there does not appear to exist a "100 % fool-proof" design for a single BMP that can achieve the entire spectrum of desirable stormwater benefits in a watershed. A clear understanding of the key mechanisms within a BMP for effluent load reductions and factors that govern these processes is a primary requirement in designing a BMP. On the same note, it must be mentioned that there has been no "exact" or "perfect" design to date. Performance variations in BMPs discussed in Chapter 2 is a fallout of these differences in design characteristics. Chapter 3 identifies the following key issues which need to be addressed in order to improve the design of stormwater BMPs:

- Influent mass loadings should be defined more clearly by considering all associated parameters that include flow rate, pollutant concentrations and their chemical forms, suspended solids and their settling velocities, dissolved solids, and the size apportionment of pollutants in the solid phase.

- Approaches to designing BMPs should focus on frequently-occurring smaller storms; the focus should be on characterizing influent load in such smaller storms and especially the parameters of concern in each watershed.
- BMP design should integrate engineering principles with hydrological characteristics, BMP performance objectives, flow attenuation, and flood control, and should incorporate design features that would enhance the BMP capability in treating the stressor(s) of concern.
- Designers should realize that one BMPs is not adequate to address the above mentioned issues and should consider the use of a treatment train; i.e., a combination of structural and nonstructural BMPs in stormwater treatment programs.

BMP Monitoring, Chapter 4, covers the complexities in developing a BMP monitoring program that yields useful results. Some difficulties and criticisms with current BMP monitoring practices are discussed in Section 4.1. This section identifies many key areas of BMP monitoring programs that require improvement in order for the data of such programs to be widely applicable. Sections 4.2 through 4.5 contain recommendations and explanations of how to ameliorate these deficiencies. Selection of appropriate parameters is covered in Section 4.2. Considerations and difficulties with monitoring nonstructural BMPs and watersheds as a whole are presented in Sections 4.3 and 4.4, respectively. Guidance on the development of a robust structural BMP effectiveness monitoring program is presented in Section 4.5. This section includes information and recommendations on planning, designing, implementing, and evaluating BMP monitoring programs. Methods for data analysis, which is particularly inconsistent between studies, is discussed in subsection 4.5.4. After reading this section, the reader should have a good handle on the problems and complexities associated with BMP effectiveness monitoring programs. The reader should also be equipped with the general knowledge and understanding necessary to develop a successful BMP monitoring program, complete with representative, quality assured, and statistically analyzed results.

BMP monitoring, especially for effectiveness, is a very complex undertaking. The number of variables that affect the resulting efficiency of a BMP is large. This, along with nonuniform sampling and analysis techniques used in current monitoring programs, has led to a wide degree of variability in reported BMP performances. The selection of appropriate pollutants is one of the most fundamental requirements in a robust BMP monitoring program. Due to the large number of variables involved in BMP performance, selecting a reasonable number of suitable parameters is difficult and requires experience and good guidance. Nonstructural BMPs have even more complications with respect to monitoring effectiveness. Without a defined influent and effluent, monitoring programs usually rely on public surveys or watershed monitoring approaches. Watershed monitoring approaches, while seemingly economical when a large number of BMPs require monitoring, is wrought with interferences from outside sources unrelated to BMPs, such as intrusion of contaminated groundwater or the inability to distinguish individual BMP performance. Thus, data from these types of monitoring programs may not be able to produce the results that shed light on the effectiveness of BMPs within the watershed.

Four steps have been outlined to monitor the effectiveness of structural BMPs. This guideline assists in the development of a robust monitoring program from planning and design phases, through the implementation and evaluation phases. A BMP monitoring program should always be initiated with clear goals and specific objectives backed by supporting background information. This foundation will minimize the risk of collecting data that is not useful. Once the goals and objectives are identified, a quality assurance project plan (QAPP) translates objectives into a plan of action. Producing a useful QAPP requires a significant amount of time upfront before any samples are taken. Although it may seem at first to be a tedious exercise, it will likely save time and money in the long run by ensuring the significance of the data collected. Design aspects, such as monitoring approach, parameter and methods selection, specifics on hydraulic, hydrologic, and water quality data collection, and methods of analysis, equipment selection, and quality assurance/quality control measures should all be clearly stated in the QAPP. Hydrologic and hydraulic (H&H) data is one of the most essential components of a well designed monitoring program. Poor quality H&H data will produce errors that will propagate through the rest of the results. Whether the QAPP calls for composite or discrete samples, they should always be flow-weighted or synchronized with flow measurements in some way. Representativeness of the collected samples is another key component that is often overlooked in BMP monitoring programs. Guidelines such as percent capture and minimum number of aliquots should be used to ensure each storm event is accurately represented. Once samples are collected, the chosen method of data analysis must provide useful and unbiased results. For example, percent removal is biased against BMPs with relatively clean influent and may not be useful for watersheds with very high influent loads as the resulting effluent, even with high percent removals, would not meet overall water quality objectives; similarly, any method that produces pollutant concentrations instead of loads will be biased against BMPs that rely on infiltration.

When BMPs are used in stormwater management, many issues need to be addressed to ensure that the BMPs are being used as effectively as possible. Chapter 5, Effective use of BMPs in Stormwater Management, covers these issues. The proper selection of a BMP is one key component to the effective use of BMPs in stormwater management. One must consider regulatory constraints, site factors, the ability of the BMP to provide stormwater quality and quantity control, cost, reliability, maintenance burden, and environmental and community acceptance. Section 5.1 walks through all these considerations and provides useful information in tabular format on many of the commonly used BMPs. Structural BMP placement is currently a “hot-button” issue. Key concerns regarding optimum and appropriate placement options are covered in section 5.2. The chapter is rounded out by a discussion of BMP integration.

An integrated approach to stormwater management appears to be the most effective use of BMPs. When multiple layers of structural and nonstructural BMPs are used in unison, the watershed will reap the largest benefit. The selection of BMPs to be used within such an integrated approach (or as single units) is dependent primarily on applicable regulations and estimated water quality and quantity performance. However, it is once again stressed that instead of relying solely on numerical efficiencies reported in the literature, a much deeper understanding of the factors that control BMP pollutant removal performance is essential to proper selection and design of a BMP. Other factors such as site characteristics, cost, reliability, maintenance requirements, and environmental and community acceptance also need consideration to ensure the chosen BMP performs as desired and expected.

BMP placement is a relatively new issue in stormwater management. Currently, political issues such as regulations that require BMPs for approval of new construction permits often control BMP placement. For this reason onsite placement is the only option. Although, onsite placement of BMPs has its advantages, more uniform sub-regional and regional BMP placement have their advantages as well. Recent efforts to identify optimal BMP placement in watersheds through modeling efforts may uphold or challenge the current focus on onsite placement practices.

Chapter 6 provides information on how to estimate the cost of structural BMPs. Costs of nonstructural BMPs are not included here as they are generally not as easily quantified as structural BMPs due to their indirect nature. Section 6.1 covers BMP cost estimating procedures and discusses four common methods of estimating costs. Section 6.2 contains information on total costs which include both capital (construction and land) and annual operation and maintenance costs. BMPs can present several tangible economic benefits in spite of their high construction (in certain cases), operation, and maintenance costs and these are discussed in Section 6.3. At other times, the use of BMPs can result in reduced infrastructure costs.

The cost of constructing any BMP is variable and can be substantial. The cost of constructing a BMP depends on many factors, including the time of year, site conditions and topography, accessibility of equipment, economics of scale, and government regulations. Several documents have been published that address cost estimation for BMPs, but most of these report only construction costs (Young *et al.*, 1996; Sample *et al.*, 2003). In addition, costs are often documented as base costs and do not include land costs, which is the largest variable influencing overall BMP cost (U.S. EPA, 1999). However, in some areas with minimum landscaping requirements, the implementation of standard practices may mean there are no "extra" land costs. The wide range of cost data reported in the literature indicates that much more information is needed in this area.

1.4 Conclusions and Recommendations

The use of BMPs to control and treat urban stormwater runoff has become a common practice in urban watershed management. This has been propagated by ordinances developed by local governments that dictate the use of structural and nonstructural BMPs for new and existing development and to protect surface water quality and mitigate the impacts of stormwater runoff on receiving waters. BMPs demonstrate a wide range of pollutant removal capabilities and their performance is affected by several factors, including the long-term variation in rainfall, BMP design characteristics, processes affecting chemical phase and speciation, and environmental conditions. The pollutant removal performance of BMPs is difficult to interpret beyond generalities due to various inter-related and complex parameters; shortcomings in current BMP related studies include:

- lack of long-term monitoring of the processes in a BMP responsible for export or detention/retention of urban stormwater pollutants,

- absence of, or inadequate monitoring within a BMP for water quality, sediment, and vegetation, which would provide a strong understanding of factors and processes that affect pollutant fate within a BMP, and
- variability in BMP performance results not only due to factors affecting the performance of BMPs, but also the methods used to characterize and calculate BMP effectiveness.

Present and future research initiatives on the implementation of structural BMPs for effective stormwater management resulting in water quality and quantity control should address the following issues:

- BMPs can be more effectively managed with a better understanding of the specific physico-chemical and biological processes and interactions that govern the transformation, immobilization and export of pollutants in stormwater. A national approach, similar to the Nationwide Urban Runoff Program (NURP), which would systemize a large number of investigations into a cohesive, well-controlled, program to learn about various BMP functions, physical mechanisms, biochemistry, and design parameters, is very much needed and in fact is underway (Urbonas, 2000). Also, the interactions among these various processes need to be understood. A knowledge of internal dynamics within a BMP, such as a pond or a wetland, is essential to improve upon BMP design and maintenance considerations.

- The type of pollutants studied should include a more extensive array of priority pollutants such as hydrocarbons, toxic inorganics and pathogens. There is less monitoring data available for pollutants such as dissolved and particulate metal species, hydrocarbons, and bacteria, (U.S. EPA, 1999). Bacteria and viruses are rarely sampled in stormwater studies, but a number of microbial pathogens can be present in stormwater and have been implicated in waterborne disease outbreaks in both humans and fish populations (Rushton, 2002). There is a need to improve existing analytical tools and establish new metrics to detect and estimate the concentrations of these pollutants. Two important research priorities include: (i) the need to develop comprehensive methods to assess the heavy metal bioavailability and other toxics following BMP treatment; and, (ii) the development of meaningful metrics to study stormwater pathogens and their ecological and health impacts. The use of conventional means of assessing water quality by only estimating the fecal coliform (FC) count is not applicable to stormwater (O'Shea and Field, 1992; Rushton, 2002). Stormwater has a wider array of pathogens including bacteria, fungi, viruses, and protozoans such as *Cryptosporidium* and *Giardia* that are not well characterized by existing methods including the FC test.

- With regard to monitoring programs, commonly used BMPs such as infiltration trenches, infiltration basins, bioretention practices, and filter strips are seldom monitored (U.S. EPA, 1999). This could be due to the difficulties associated with collecting inflow and outflow samples from these systems. These systems are widely used, yet their effectiveness has not been well documented; pilot-scale research investigations would be a good approach to assess their effectiveness.

- BMP monitoring should be conducted over a relatively long time period (one yr or more) continuously during dry- and wet-weather flow conditions using an influent-effluent mass balance approach. BMP monitoring and data reporting should be more frequent, (e.g., on a monthly and seasonal basis) rather than estimating annual averages of pollutant removal. This temporal scale approach helps to better assess seasonal factors such as thermal stratification of ponds, plant growth and senescence that influence the performance of BMPs. Also, the investigation of dry- weather/low- flow samples would be the key to ascertain if pollutants are released under low flow conditions.

- The current approach in the use of urban BMPs perceives the effect these practices would have just on water quality. BMPs are best characterized by: (i) how much runoff is prevented; (ii) volume of runoff being treated; and, (iii) the quality of the resulting effluent. In addition, BMP objectives should foremost extend beyond water quality and take into account other media including sediments, vegetation, and benthic invertebrates. The role of vegetation, bacteria, and benthic invertebrates in the accumulation and/or breakdown of pollutants need to be more intensively investigated.

- Most of the commonly used structural BMPs extensively rely on sedimentation as the predominant mechanism to remove pollutants from the water column. What is largely ignored is that these contaminated sediments should be subject to appropriate maintenance practices; they have the potential to either leach the organic and inorganic contaminants into the ground below, or may result in pollutant resuspension during high flows or under unfavorable environmental conditions, which include changes in pH and/or the oxidation reduction potential (ORP). Any BMP research or monitoring program should have in its analytical scheme the sediment speciation for contaminant assessment in order to understand the bioavailability and the potential for pollutant resuspension in the water column.

- A large number of BMP studies at present focus on seasonal, short-term monitoring activities; the long-term performance of BMPs (>5 yr) is uncertain. The pollutant removal capabilities of BMPs are likely limited by a finite capacity of sediment/substrate to sorb and retain pollutants. Numerous research findings infer that the longevity of BMPs is linked to the ability of the substrate to assimilate pollutants and maintenance practices (Schueler *et al.*, 1992), lending credence to the need for an evaluation of the long-term performance of these systems.

REFERENCES

State of California Department of Transportation (Caltrans). (2002). *Stormwater Quality Handbooks Project Planning and Design Guide.* http://www.dot.ca.gov/hq/oppd/stormwtr/PPDG-stormwater-2002.pdf.

D'Arcy, B. and A. Frost. (2001). The role of best management practices in alleviating water quality problems associated with diffuse pollution. *The Science of The Total Environment* **265**: 359-367.

Friends of the Lower Olentangy Watershed (FLOW). (2003). *The Lower Olentangy Watershed*

Action Plan in 2003 - Strategies for Protecting and Improving Water Quality and Recreational Use of the Olentangy River and Tributary Streams in Delaware and Franklin Counties. Columbus, OH. http://www.olentangywatershed.org/LowerOlentangyActionPlan04.pdf.

Leeds, R., L.C. Brown, and N.L. Watermeier. (1993). *Nonpoint Source Pollution: Water Primer. AEX-465-93. Ohio State University Extension Fact Sheet*. Food, Agricultural and Biological Engineering, The State University of Ohio, Columbus, OH. http://ohioline.osu.edu/aex-fact/0465.html

Massachussets Department of Environmental Protection and Massachussets Office of Coastal Zone Management (MA DEP and MA CZM). (1997). *Stormwater Management Volume Two: Stormwater Technical Handbook*. Boston, MA.

Rushton, B. (2002). BMP monitoring: methods and evaluations, http://www.stormwater-resources.com/library.htm#BMPs (#147PL). Stormwater News, Cocoa Beach, Florida.

Sample, D.J., J.P. Heaney, L.T. Wright, C.Y. Fan, F.H. Lai, and R. Field. (2003). Cost of best management practices and associated land for urban stormwater control. *Journal of Water Resources Planning and Management* **129**(1): 59-68.

Schueler, T.R. (1987). *Controlling Urban Runoff: A practical manual for planning and designing urban BMPs*. Metropolitan Washington Council of Governments, Washington, DC.

Schueler, T.R., P.A. Kumble, and M.A. Hearty. (1992). *A Current Assessment of Urban Best Management Practices: Techniques for reducing nonpoint source pollution in the coastal zone*. Metropolitan Washington Council of Governments, Washington, DC.

United States Environmental Protection Agency (U.S. EPA). (1993). *Handbook - Urban Runoff Pollution Prevention and Control Planning. EPA-625-R-93-004*. Washington, DC.

United States Environmental Protection Agency (U.S. EPA). (1999). *Preliminary Data Summary of Urban Stormwater Best Management Practices. EPA-821-R-99-012*. Washington, DC.

United States Environmental Protection Agency (U.S. EPA). (2000). *National Water Quality Inventory 1998 Report to Congress. EPA-841-R-00-001*. Washington, DC.

United States Environmental Protection Agency (U. S. EPA). (2001). National Menu of Best Management Practices for Stormwater Phase II, http://cfpub.epa.gov/npdes/stormwater/menuofbmps/menu.cfm. Washington, DC.

Urbonas, B. (2000). Assessment of stormwater best management practice effectiveness (chapter 7). In: *Innovative urban wet-weather flow management systems*, Field, R., J.P. Heaney, and R. Pitt, (editors). Technomic Publishing Co., Inc., Lancaster, PA. Pages 255 - 300.

Young, G.K., S. Stein, P. Cole, T. Kammer, F. Graziano, and F. Bank. (1996). *Evaluation and Management of Highway Runoff: Water Quality Technical Report*. Department of Environmental Programs, Metropolitan Washington Council of Governments, Washington, DC.

2 Types of Best Management Practices

Swarna Muthukrishnan, Richard Field and Daniel Sullivan

CONTENTS

2.0 INTRODUCTION

A stormwater best management practice (BMP) is a technique, measure, or structural control that is used to manage the quantity and improve the quality of stormwater runoff in the most cost effective manner. The U.S. EPA (1999b) defines best management practices as "schedules of activities, prohibitions of practices, maintenance procedures, and other management practices to prevent or reduce the pollution of waters of the United States." BMPs also include, but are not limited to, treatment requirements, operating procedures, and practices to control plant site runoff, spillage or leaks, sludge or wastewater disposal, or drainage from raw material storage." There are two general types of BMPs used to reduce the threat of stormwater runoff pollution from construction and development in urbanizing areas: (i) Nonstructural or source control BMPs and (ii) Structural or treatment BMPs (U.S. EPA, 1993b).

Nonstructural BMPs refer to those stormwater runoff management techniques that use natural measures to reduce pollution levels, do not require extensive construction efforts, and either limit the generation of stormwater runoff, or reduce the amounts of pollutants contained in the runoff. They do not involve fixed, permanent facilities and they usually work by changing behavior through government regulation (e.g., planning and environmental laws), persuasion, and/or economic instruments (Taylor and Wong, 2002). These BMPs include institutional, educational or pollution prevention practices. Because they improve runoff quality by reducing the use, generation and accumulation of potential stormwater contaminants at or near their sources in many cases, they are also termed as source control BMPs (WEF and ASCE, 1998; U.S. EPA, 1999b).

Structural BMPs are engineered systems and methods designed to provide temporary storage and treatment of stormwater runoff for the removal of pollutants (MWLAP, 1992; MDE, 2000; Clar *et al.*, 2003). These practices are aimed at "controlling the volume and discharge rate of stormwater runoff from urban areas, as well as, reducing the magnitude of pollutants in the discharge water through rotation, physical containment or flow restrictions designed to allow settling, physical removal through filtration, percolation, chemical precipitation or flocculation, and/or biological uptake and may be called structural or treatment BMPs" (Florida DER, 1988). Structural BMPs improve the quality and/or control the quantity of stormwater runoff; common examples include detention ponds and constructed wetlands.

The distinction between nonstructural and structural BMPs is very clear in some cases, but less so in others. For example, street sweeping for pollutant removal is one BMP that could be considered either a structural or nonstructural control (WEF and ASCE, 1998). The use of vegetation to disconnect directly impervious surfaces such as rooftops, driveways, parking lots and streets, is another example of a BMP that could be considered as a structural or nonstructural control. Some of the newer concepts for urban stormwater runoff management such as better site planning techniques (CWP, 1998) and low-impact development (LID) technologies such as "green roofs" (U.S. EPA, 2000a, 2000b) focus on the use of both planning techniques and micro-scale integrated landscape-based practices to prevent or reduce the impacts of urban stormwater runoff at the very point where these impacts would be initially generated or just downstream. These approaches tend to have a great deal of overlap between preventative source control approaches and small-scale treatment approaches that blur the distinction between these two types of BMPs. It is being recognized that a combination of nonstructural controls and one or more structural or treatment BMPs, often referred to as a "treatment train approach," may be needed to meet water quality objectives, depending on the stormwater management goals and objectives identified for a specific site or area. For example, if the pollutants of concern are a combination of trash/debris, dissolved copper and phosphorus, it is more likely that a treatment train would be required to be successful in treating the stormwater runoff.

The list of BMPs, both structural and nonstructural, presented in the following discussion is based on information available at the time that this book was written. The development of stormwater

BMPs is an emerging science and stormwater treatment technologies are frequently being modified; also, efforts are increasingly directed towards pollution prevention and the implementation of nonstructural practices. Finally it is being increasingly recognized that given the number and forms of the pollutants of concern, multiple treatment processes are required. Keeping this in mind, the list of BMPs is provided in this document as a guide only and is not meant to be exclusive of alternative stormwater BMPs that could be used in urban watershed management.

2.1 NONSTRUCTURAL OR SOURCE CONTROL BMPs

Nonstructural or source control BMPs are practices that prevent pollution by reducing potential pollutants at their source before they come into contact with stormwater, or capturing and disposing of stormwater at its source. These BMPs aim to eliminate contamination by preventing their introduction into the environment. Often termed pollution prevention practices, these BMPs can include any method that avoids or reduces potential exposure of contamination to the elements. Unfortunately, nonstructural BMPs are often not being given due credit in watershed management plans (Clar *et al.*, 2003) as data on their performance is virtually not available. However, there is an increasing recognition of the primary need for pollution prevention rather than treatment of polluted stormwater in long-term urban watershed management program implementation. There are two major reasons for this shift in the approach to stormwater management:

- Stormwater management approaches are increasingly being directed toward the development of least-cost measures to treat stormwater pollution, which include nonstructural and low-cost structural controls.
- Nonstructural BMPs are found to be very effective in controlling pollution generation at the source and can eliminate or reduce the need for costly end-of-pipe treatment by structural BMPs (WEF and ASCE, 1998).

Nonstructural controls include regulatory controls that prevent pollution problems by controlling land development and land use, as well as source controls that reduce pollutant buildup or lessen its availability for washoff during rainfall (City of Austin, 1988; MWLAP, 1992; U.S. EPA, 2000a, 2000b). Nonstructural practices can play a significant role in reducing water quality impacts; they are increasingly being recognized as a critical feature of every stormwater BMP plan, especially with regard to site design and development and maintenance costs. The key benefit of some nonstructural practices is that they can also reduce the generation of stormwater runoff from the site, thereby reducing the size and cost of storage; in addition to providing partial removal of many pollutants (MDE, 2000). A comprehensive plan addressing urban stormwater runoff pollution prevention and control requires the implementation of a combination of nonstructural and structural BMPs for existing and new development.

Nonstructural or source control approaches for stormwater management comprise three major components:

- planning, design and construction of developments and re-developments to minimize or eliminate adverse impacts;

- good maintenance of impervious and pervious surfaces to minimize exposure and release of pollutants; and
- education and training to promote awareness of the potential problems associated with urban stormwater runoff and of specific BMPs to help solve problems (WEF and ASCE, 1998).

Typically, nonstructural BMPs for urban areas can be grouped into categories as shown in Table 2-1 (WEF and ASCE, 1998).

Table 2-1. Nonstructural BMPs for Urban Stormwater Runoff

Major Categories	Nonstructural Practice
2.1.1 Public Education	▸ Public Education and Outreach
2.1.2 Planning and Management	▸ Better Site Design ▸ Vegetation Controls ▸ Reduction/Disconnection of Impervious Areas ▸ Green Roofs* ▸ Low-Impact Development**
2.1.3 Materials Management	▸ Alternative Product Substitution ▸ Housekeeping Practices
2.1.4 Street/Storm Drain Maintenance	▸ Street Cleaning ▸ Catchbasin Cleaning ▸ Storm Drain Flushing ▸ Road and Bridge Maintenance ▸ BMP Maintenance ▸ Storm Channel and Creek Maintenance
2.1.5 Spill Prevention and Cleanup	▸ Above Ground Tank Spill Control ▸ Vehicle Spill Control
2.1.6 Illegal Dumping Controls	▸ Illegal Dumping Controls ▸ Storm Drain Stenciling ▸ Household Hazardous Waste Collection ▸ Used Oil Recycling
2.1.7 Illicit Connection Control	▸ Illicit Connection Prevention ▸ Illicit Connection - Detection and Removal ▸ Leaking Sanitary Sewer and Septic Tank Control
2.1.8 Stormwater Reuse	▸ Landscape Watering ▸ Toilet Flushing ▸ Cooling Water ▸ Aesthetic and Recreational Ponds

* Also considered a structural BMP since engineering principles and design are involved
**Considered a combination of both nonstructural and structural BMP
(WEF and ASCE, 1998)

2.1.1 Public Education

Public understanding, involvement and support is an essential component of stormwater management programs. A public education and participation plan provides the municipality with a strategy for involving the public in making stormwater management decisions, and for educating its employees, the public and businesses about the importance of protecting stormwater from improper use, storage and disposal of pollutants. The intent of programs targeting public involvement is to get the public behind the overall stormwater management program to increase its acceptance. These programs are designed to raise public awareness on diffuse source pollution and encourage people to change their everyday behavior in the use and disposal of chemicals, other household and automotive products in order to minimize and/or prevent the entry of these products into stormwater runoff. Essentially, such programs include conducting workshops, open houses, surveys and other means of including the public in providing input into and commenting on stormwater management efforts. This may be accomplished by distributing educational materials to the community or by conducting equivalent outreach activities about the impacts of stormwater discharges on water bodies and the steps the public can take to reduce pollutants in urban stormwater runoff (U.S. EPA, 2001). This promises to be a cost-effective way of stormwater quality management; yet, its effectiveness on the actual reductions of target constituents in receiving waters has yet to be definitively demonstrated (WEF and ASCE, 1998).

The objectives of a public education and participation plan should be to:

- promote clear identification and understanding of the problem and the solutions,
- identify responsible parties and efforts to date,
- promote community ownership of the problem and its solutions,
- change behaviors, and
- integrate public feedback into program implementation.

The main components of a public education program include the following (U.S. EPA, 1999b).

- Public outreach/education for homeowners
  - Lawn and garden activities education
  - Water conservation practices for homeowners
  - Pet waste management education
  - Proper disposal of household hazardous wastes
  - Trash management education
- Targeting public outreach/education
- Public outreach programs for new development
- Pollution prevention programs for existing development
-

2.1.2 Planning and Management of Developing Areas

These practices by local governments and large land owners/developers are aimed at reducing urban stormwater runoff and the discharge of pollutants through stormwater from new developments, and are most effective when applied during the site-planning phase of new development. This BMP presents an important opportunity to reduce the pollutants in stormwater runoff by using a comprehensive planning process to control or prevent certain land use activities in areas where water quality is sensitive to development. It is applicable to all types of land use and represents one of the most effective pollution prevention practices. Land use planning and management are critical to watershed management.

2.1.2.1 Better Site Design

The use of better site design techniques is one of the few watershed management practices that seeks to simultaneously reduce pollutant loads, conserve natural areas, save money and increase property values, and at the same time it collectively employs a variety of methods to accomplish three goals at every development site, to: (i) reduce the amount of impervious cover and/or directly connect impervious cover; (ii) increase natural lands set aside for conservation; and, (iii) use pervious areas for more effective stormwater treatment (CWP, 2000). However, in many communities, it is observed that many of the regulations governing the development process in such areas as zoning, parking and street standards, and drainage, to name a few, are at cross-purposes with better site design and need to be addressed. Better site design practices identify areas where existing codes and standards can be changed to better protect streams, lakes, and wetlands at the local level, and fall into the following three categories.

- narrow residential streets and smaller parking lots such as minimizing residential street cul-de-sac sizes, reducing overall imperviousness associated with parking lots;
- lot development such as open space design development, routing of rooftop runoff to pervious areas; and
- conservation of natural areas such as buffer zones, setbacks and easements.

2.1.2.2 Vegetation Control

This typically involves a combination of mechanical methods and careful application of chemicals (fertilizers, insecticides and herbicides). Mechanical vegetation controls include practices such as leaving existing vegetation, less frequent cutting, hand cutting, planting low-maintenance vegetation, collecting and properly disposing of clippings and cuttings, and educating the public and employees. Stormwater quality strongly impacts areas including steep slopes, vegetated drainage channels, creeks, areas adjacent to catch basins, detention/retention basins. Flat or relatively flat vegetated areas, areas not adjacent to drainage structures, and areas screened from drainage structures or from vegetation are less impacted (WEF and ASCE, 1998).

2.1.2.3 Reduction/Disconnection of Impervious Areas

The volume of stormwater runoff generated in a development can be greatly reduced by minimizing the amount of impervious surfaces. Reductions in impervious area can be undertaken by reducing the overall size of the developed area, and/or by reducing the amount of impervious surface created within the developed area. Disconnection of impervious surfaces can be undertaken by directing runoff from roofs and paved surfaces over vegetated surfaces before it reaches the drainage conveyance system (GVSDD, 1999). Reductions in impervious area can also be achieved through cluster developments that maximize open (undeveloped) space and minimize the required length of roadway and other infrastructure. Clustering concentrates development on smaller lots into compact areas and leaves relatively large areas undeveloped, in contrast to conventional grid developments that cover the entire site with larger lots and result in more overall impervious area (roads, sidewalks etc). This approach will help address peak flow control, stream bank erosion protection, removal of drainage path obstruction, water quality enhancement, groundwater recharge and community enhancement. Clustering is suitable for new developments and redevelopment areas, but has limited opportunities for use in areas of existing development. A few measures that reduce impervious surfaces include:

- reducing the width of streets,
- limiting the length and radius of cul-de-sacs,
- using porous pavement of modular block pavers in parking areas and low-traffic areas,
- placing sidewalks to only one side of the street or no sidewalks where possible, and
- reducing frontage requirements to lessen paved surface areas (U.S. EPA, 1999b).

2.1.2.4 Green Roofs

Green roofs, also known as vegetated roof covers, eco-roofs or nature roofs, are multi-beneficial structural components that help to mitigate the effects of urbanization on water quality by filtering, absorbing or detaining rainfall (GreenRoofs; Liptan and Strecker, 2003). In areas of high-density development, where pervious surfaces and open ground often make up 10% or less of total surface area capable of absorbing or diverting storm water runoff, green roofs are one of the best ways to reduce runoff volumes via evapotranspiration losses. In addition, they can provide significant social, environmental and financial benefits. There has been a lot of discussion with due explanation, about the need to describe green rooftops as a structural or a nonstructural BMP; this chapter examines green roofs as a nonstructural BMP, mainly for the reason that the goal of green rooftops is runoff volume reduction, pollution prevention and mitigation at the source (hydrological source control) with minimal and cost-effective landscape features located at the lot level, as opposed to structural controls that convey, manage and treat stormwater in large, expensive, end-of-pipe facilities.

Green roof systems involve the creation of a "contained" green space atop a human-made roofing system that include the roof structure and insulation; a waterproofing membrane, often with root repellent; a drainage layer; landscape or filter cloth to contain roots and soil; a specialized growing medium (which may not include soil); and plants (MWLAP, 1992). They comprise an impermeable membrane or similar structure that supports a lightweight soil medium and living vegetation, e.g., grass or groundcover placed on all or part of the building roof (Metro Council, 2001). The soil and vegetation layer provide a means of replacing the impermeable surfaces of building roofs to reduce stormwater runoff volumes, control stormwater peak flows, improve stormwater quality, and reduce stormwater runoff temperature. The soil is planted with a specialized mix of plants that can thrive in the harsh, dry, high temperature conditions of the roof and tolerate short periods of inundation from storm events. Care should be exercised about the soils employed to be sure that they do not become a source of pollutants, for example, nutrients.

Historically, engineered green roofs originated in northern Europe, where sod roofs and walls have been utilized as construction materials for hundreds of years. The development of contemporary approaches to green roof technology began in the urban areas of Germany over 30 years ago. Despite centuries of use in Iceland and various research initiatives to encourage their use in Canada, Germany, France, Austria and Switzerland, green roof tops are only a recent entry in the U.S. However, several new and planned projects, including a retrofit of the Chicago City Hall, a 450,000 ft^2 vegetated roof planned for a new Ford assembly plant in Dearborn, MI, a vegetated roof at the Green Institute in Minneapolis, a vegetated roof top at the Fencing Academy, Philadelphia (U.S. EPA, 2000c), numerous green roofs in Portland, OR, encouraged by a city incentive program (Liptan and Strecker, 2003), and the Green Roof research center at the University of Pennsylvania dedicated to demonstrating and promoting green roof research, education and technology transfer, are raising visibility and encouraging the nascent U.S. market for green roof technology.

Green roofs are especially effective in controlling intense, short-duration summer storms and have been shown to reduce cumulative annual runoff by 50% in temperate climates with summer storm reductions of up to 80 or 90%. Green roof systems partially restore the natural hydrologic cycle, compensating for the impervious surface areas associated with urban development. These systems absorb and retain precipitation; the water is used by the plants or slowly released to the storm system. In general, green roof systems provide summer retention rates of 70 to 100% and winter rates of 40 to 50% depending on factors such as precipitation patterns and forms, substrate, type and depth of vegetation, temperature, humidity, sun and wind. In addition, they moderate the temperature of runoff and filter nutrients, sediments and metals from the runoff.

Green roofs provide stormwater management benefits by:

- utilizing the biological, physical, and chemical processes found in the plant and soil complex to prevent airborne and rain-entrained pollutants from entering the storm drain system; and
- reducing the runoff volume and peak discharge rate by holding back and slowing down the water that would otherwise flow quickly into the storm drain system.

The hydrologic processes that can be influenced by design choices and aid in managing stormwater include: interception of rainfall by foliage and subsequent evaporation; soil moisture penetration, soil adsorbing ability, and soil moisture maintenance (e.g., irrigation); reduction in the velocity of runoff; shallow subterranean flow through the soil; root zone moisture uptake; and, evapotranspiration (U.S. EPA, 2000a).

2.1.2.5 Low-Impact Development (LID)

LID is a site design strategy with the goal of maintaining or replicating or minimizing the change in the pre-development hydrologic regime through the use of design techniques to create functionally equivalent hydrologic landscapes. Hydrologic functions of storage, infiltration, and ground water recharge, as well as the volume and frequency of discharges are maintained, or their changes reduced through the use of integrated and distributed micro-scale stormwater retention and detention areas, reduction of impervious surfaces, and the lengthening of flow paths and runoff time (U.S. EPA, 2000d). Over the last several years, a number of techniques when applied in combination in a methodological way, have asserted to result in little or no stormwater impacts. These techniques began to be increasingly implemented in stormwater management efforts in the early 1990's and were given the term Low-Impact Development (LID) by Prince George's County (Strecker, 2001).

The principles of LID are based on controlling stormwater at the source by the use of micro scale controls that are distributed throughout the site. These are multi functional site design practices and incorporate alternative stormwater management practices such as functional landscapes that act as stormwater control facilities, depression storage and open drainage swales. Such systems of controls can reduce or eliminate the need for a centralized BMP facility for the control of stormwater runoff. LID measures provide a means to address both pollutant removal and the protection of pre-development hydrologic functions. Some basic LID principles include conservation of natural features, minimization of impervious surfaces, hydraulic disconnects, disbursement of runoff, and phytoremediation. LID practices such as bioretention facilities or rain gardens, grass swales and channels, vegetated rooftops, rain barrels, cisterns, vegetated filter strips and permeable pavements perform both runoff volume reduction and pollutant filtering functions (Prince George's County (DER), 2002). The benefits of LID are many:

- LID addresses hydrologic changes caused by development at the site level, which reduces the downstream impact of increased imperviousness.
- LID practices, when used in combination with each other and with traditional treatment practices such as regional retention ponds, further reduce pollutant loading to receiving water bodies.
- Many LID practices involve natural landscaping, including the planting of trees, shrubs, and flower gardens that enhance site aesthetics and reduce mowing requirements; the choice of local species and locally grown stocks can greatly reduce watering and fertilizer requirements.
- Careful regrading and well-sited localized depression and storage areas with slower but assured drainage help improve overall site drainage, prevent overly long pooling and creation of mosquito-breeding habitat, and reduce both onsite and downstream flooding.

2.1.3 Materials Management

The responsible management of common chemicals, such as fertilizers, solvents, paints, cleaners, and automotive products can significantly reduce polluted stormwater runoff (U.S. EPA, 2001). Such products must be handled properly in all stages of their useful lives. These practices include controlling the use, storage and disposal of chemicals that could pollute stormwater runoff with the objective of reducing the opportunity for rainfall or runoff to come into contact with these chemicals. This BMP includes activities pertaining to material use controls, material exposure controls, and material disposal and recycling controls. Material storage controls help prevent or reduce the discharge of pollutants to stormwater from material delivery and storage areas for municipal and commercial operations. This BMP primarily concerns the design and maintenance of material storage areas that reduce exposure to stormwater by storing materials inside or under cover on paved surfaces; using secondary containment, where required; minimizing storage and handling of hazardous materials; and, regularly inspecting storage areas (WEF and ASCE, 1998).

Materials management primarily includes use minimization and the use of less toxic alternatives and recommends the following approaches as control measures:

- using products only when needed,
- using chemicals according to label instructions or lesser quantities, and
- applying only when required.

2.1.3.1 Alternative Product Substitution

The presence of toxic substances in stormwater and receiving waters can be drastically reduced by prudent use of alternatives to toxic substances. The most common toxic substances found in homes and small businesses include cleaners, automotive products, pesticides, fertilizers, paints and fuels that find their way into ground and surface receiving waters upon improper disposal. Alternatives exist for most product classes; some examples include rechargeable batteries, baking soda, pump-type or non-aerosols etc. Most garden products have natural alternatives, and less toxic alternatives exist for home and automotive repair products (WA DOE, 2001). For example, alternatives to pesticides, such as insecticidal soap or natural bacteria recommended in integrated pest management programs can help reduce the need for pesticides (U.S. EPA, 1999b). Integrated pest management strategies include the use of natural predators and pathogens, mechanical controls, native and pest-resistant plants, removing pest habitat and localized use of appropriate use of chemicals as a last resort when problems are observed (Taylor and Wong, 2002).

Care should also be exercised in selecting building materials for new construction and renovation. Commonly used building materials such as zinc roof gutters and architectural copper can leach significant quantities of dissolved metals into runoff causing adverse ecological impacts. The use of proper alternative materials can reduce potential contaminants in runoff by eliminating compounds that leach into runoff. This will reduce the need for pesticide application, painting and other maintenance by reducing the volume of runoff. Alternative building materials are available instead of treated copper base (or other treatment of wood) as lumber for decking (plastic or plastic-wood composites), roofing materials (coated metal products, roofing materials made of recycled rubber or plastic instead of metals), home siding (vinyl, cement-fiber), and paving for driveways, decks and sidewalks (porous concrete and asphalt, modular blocks and crushed granite). Other examples include the use of formaldehyde-free plywood, and low VOC paints, carpets, and pads (CASQA, 2003).

The promotion of safer alternative products, however, should be coupled with other programs designed to reduce the presence of hazardous or toxic materials in homes, offices, commercial spaces and more importantly, urban stormwater runoff. Examples of such programs are hazardous materials collection, good housekeeping or material management practices, oil and automotive waste recycling, and spill response and prevention (WEF and ASCE, 1998). Examples of proactive communities (those that have instituted effective programs promoting safer alternative products) include Santa Clara County, the city of Palo Alto, the city and county of San Francisco, CA, and the cities of Bellevue and Seattle, WA.

2.1.3.2 Housekeeping Practices

The promotion of efficient and safe housekeeping practices (which include storage, use, cleanup, and disposal) when handling potentially harmful materials such as pesticides, fertilizers, cleaning solutions, paint products, automotive products, and swimming pool chemicals can be an effective source control BMP. The target audiences for making this BMP an effective one are municipal employees, the general public and small and large businesses. This BMP should be implemented in conjunction with safer alternative products BMPs and integrated to the highest extent possible with existing programs in the municipality.

Commercial and retail areas can contribute significantly to pollutant loadings in urban stormwater runoff, with the largest contribution from impervious surfaces used for vehicle parking, storage and maintenance areas that contribute metals, sediments and hydrocarbons. Good housekeeping practices include using porous pavement or modular paving systems for vehicle parking lots, minimizing or eliminating exposure of materials and equipment to rainfall by storing inside or under cover, using dry cleanup techniques instead of wet techniques, minimizing pesticide/herbicide and fertilizer use, limiting discharges of equipment wash water to storm drains, and limiting direct runoff of rooftops to storm drains (U.S. EPA, 1993a, 1999b).

2.1.4 Street/Storm Drain Maintenance

This BMP category applies to the removal of pollutants from paved areas and the maintenance of runoff quality controls that exist within the drainage system. Examples include street and parking lot sweeping, catch basin cleaning, road and bridge maintenance, and maintenance of structural controls in the system for runoff quality management. This group also includes the use of good housekeeping measures whenever performing pavement maintenance such as installing asphalt overlays or conducting seal and chip procedures.

2.1.4.1 Street Cleaning

Street and parking area cleaning was extensively studied during early U.S. EPA-funded research projects since the 1970s, as it was thought to be an effective means of controlling the quality of runoff associated with large quantities of pollutants found on streets and parking lots (Field and Sullivan, 2003). This management measure involves employing pavement cleaning practices such as street sweeping on a regular basis to minimize pollutant export to receiving waters. These cleaning practices are designed to remove from road and parking lot surfaces sediment debris and other pollutants that are potential sources of pollution affecting urban waterways (FHWA, 2000). Street sweeping is considered to be an effective ultra-urban best management practice for reducing total suspended solids and associated pollutant washoff from urban streets. This is especially well suited to those urban environments with little land available for the installation of structural controls. Areas of application include commercial business districts, industrial sites, and intensely developed areas in close proximity to receiving waters. In highway applications, street-sweeping may be considered for road shoulders (where safety permits), rest stop parking areas, or maintenance yards.

There are many kinds of street sweepers currently being used: mechanical; vacuum-assisted sweepers; tandem sweeping; regenerative air sweepers; and, vacuum-assisted dry sweepers.

Performance evaluation of street sweeping as a stormwater quality BMP under the NURP Program (U.S. EPA, 1983) concluded that while this technique was effective in removing litter and coarse fractions of sediment from the surface, it provided no significant reduction in nutrient concentrations in stormwater runoff. However, newer street sweeping technology more effectively removes the finer fraction of suspended particles that carry a substantial portion of the stormwater pollutant load; sweeping programs carried out routinely on a weekly or bimonthly basis have shown potential reductions of up to 80% in annual loads of total suspended solids and associated pollutants (FHWA, 2000; U.S. EPA, 2001).

Although street sweeping has been widely practiced for litter and dust control, its implementation as a stormwater pollution control practice is a fairly recent development (NVPDC, 1996). "The effectiveness of street sweeping programs depends more on factors such as land-use activities, the inter-event dry period, street sweeping frequency and timing, access to source areas and sweep operation than the actual street sweeping mechanism" (Walker and Wong, 1999, in Taylor and Wong, 2002). The performance of street sweepers is significantly affected by factors that include street dirt loadings, street texture, litter and moisture, parked car conditions, and equipment operating conditions. It must be noted that extensive street cleaning might be more beneficial in reducing SS and heavy metals in arid western regions of the U.S., where the infrequent rains allow substantial pollutant accumulation on the street; however, in the wet southeastern U.S. where large and frequent rains occur, "street cleaning is likely to have much less direct water quality benefit, beyond possibly important litter and floatable control" (Field and Sullivan, 2003).

2.1.4.2 Catchbasin Cleaning

Catchbasins are chambers or sumps, usually built at the curb line, that allow surface water runoff to enter the stormwater conveyance system. Many catchbasins have a low area below the invert of the outlet pipe intended to retain coarse sediment. Catchbasins as BMPs are reasonably effective in protecting sewers from receiving loads of very coarse solids greater than 0.04 in. diameter, but not very effective in capturing fine particulates such as clays or silt (MDEQ, 1998). Catchbasins naturally accumulate sediment and debris such as trash and leaf litter; sediment trapping by catch basins prevents solids from clogging the storm sewers and being washed into receiving waters. Uncleaned catchbasins which allow coarse materials to overflow may contribute higher loads of BOD (biochemical oxygen demand) and sediment to the receiving streams. Catchbasins in order to be effective require periodic cleaning using either a vacuum or adductor to remove the accumulated pollutants and maintain the pollutant removal efficiency (U.S. EPA, 1999a, 1999b). One recommendation is to clean the catchbasins at least two times per yr (just before and after the rainy season), and/or when the catchbasin storage is one-third full, whichever happens first, as this will help keep pollutants and sediments from re-entering stormwater (Pierce County). Some limitations associated with catchbasin cleaning include:

- Catchbasin debris usually contain appreciable amounts of water and "offensive" (odorous) organic matter that must be properly disposed of.

- Catchbasins may be difficult to clean in areas with poor accessibility, traffic congestion and parking problems.
- Cleaning is difficult in winter in the presence of snow and ice (U.S. EPA, 1999a).

Sediment and debris removed from catchbasins can potentially be classified as hazardous waste particularly when cleaning is infrequent and/or in industrial areas and always needs to be disposed of properly to avoid negative environmental impacts.

2.1.4.3 Roadway and Bridge Maintenance

Road and street surfaces undergo breakdown due to frictional action of traffic, freeze-thaw temperatures, frost heaving, ultraviolet degradation and erosion of road subbase. This results in exposure of unstabilized subbase material to erosive forces of water and subsequent increases in suspended solids concentration as well as other constituents such as PAHs. The substantial loadings of sediments and other pollutants generated during daily roadway and bridge use and scheduled repair operations pose a threat to local water quality by contributing heavy metals, hydrocarbons, PAHs, sediment and debris to stormwater runoff (U.S. EPA, 2001). A few measures that help to alleviate the impact of pollutants from roadway surfaces include:

- Routine performance of maintenance activities such as sweeping, vegetation maintenance, and cleaning of runoff control structures.
- Modifications in roadway resurfacing practices and application techniques for salt and other deicers.

Extensive studies on the characterization and environmental impacts of highway deicing chemicals have been conducted by the U.S. EPA and the negative environmental impacts of these salting and sanding operations is well documented (D'Itri, 1992). Alternative de-icing products such as acetates, formates and agricultural residues can be used if impacts due to traditional deicing products are significant (U.S. EPA, 1999a, 1999b)

2.1.4.4 Storm Drain Flushing

Overflowing of storm drains and storm drain inlets, as well as the resulting increase in erosion are natural consequences of irregular cleaning and maintenance of these facilities. Routine cleaning of storm drains is associated with a lot of benefits, including increased dissolved oxygen, reduced levels of bacteria and support of instream habitat. Flushing a storm drain with water to suspend and remove deposited materials is particularly beneficial for storm drain pipes with relatively flat grades or low flows where self-cleansing becomes difficult (U.S. EPA, 2001). Flushing helps to ensure that pipes convey design flow and removes pollutants from the storm drain system (WEF and ASCE, 1998).

2.1.4.5 BMP Maintenance

BMPs require a variety of periodic maintenance activities in order to enhance performance. This includes sediment removal, vegetation maintenance, periodic maintenance and repair of outlet structures if needed, periodic replacement of filter media, to name a few (U.S. EPA, 1999b). Regular inspection of control measures is essential in order to maintain the effectiveness of post-construction stormwater BMPs. The inspection and maintenance of BMPs can be categorized into two groups: expected routine maintenance, and non-routine (repair) maintenance. Routine maintenance involves checks performed on a regular basis to keep the BMP in good working order and aesthetically pleasing and is an efficient way to avoid the health and safety threat inherent in BMP neglect (e.g., prevent potential nuisance situations, reduce the need for repair maintenance, reduce the chance of polluting stormwater runoff by finding and correcting problems before the next rain) (U.S. EPA, 2001). A general guideline on BMP maintenance is presented in more detail in Chapter 5 of this book.

2.1.4.6 Storm Channel and Creek Maintenance

Regular removal of illegally dumped items and material from storm drainage channels and creeks helps to reduce pollutant levels in stormwater runoff. The approaches for effective storm channel and creek maintenance include the following:

- identify and regularly clean up stormwater "hotspots" and other storm drainage areas where illegal dumping and disposal occurs,
- establish and maintain buffer zones along creeks, and
- modify storm channel characteristics: e.g., improve channel hydraulics; reduce channel erosion and increase pollutant removals; and enhance aesthetics and habitat values.

2.1.4.7 Stormwater "Hotspots"

Stormwater "hotspots" are areas of the urban landscape that often generate higher concentrations and/or loads of certain pollutants, such as hydrocarbons, trace metals or toxicants than are normally found in urban stormwater runoff and that are termed so based on monitoring studies (MDE, 2000; Atlanta Regional Commission, 2001). These areas merit special management and the use of specific pollution prevention activities and/or structural stormwater controls. These are areas where land use or activities generate potentially highly contaminated stormwater runoff, with concentrations of pollutants in excess of those typically found in urban stormwater.

The designation of a site as a stormwater hotspot has important implications for how the stormwater is managed, the most important one being that untreated stormwater runoff from hotspots cannot be allowed to infiltrate into groundwater where it may contaminate water supplies. Secondly, a higher level of stormwater treatment is needed at hotspot sites to prevent pollutant washoff after construction. This typically involves preparing and implementing a stormwater pollution prevention plan that involves a series of operational practices at the site to reduce the generation of pollutants by preventing contact with rainfall. The following land uses and activities are typically considered as stormwater hotspots when exposed to stormwater (VA DCR, 1999; Atlanta Regional Commission, 2001; CASQA, 2003):

- vehicle salvage yards and recycling facilities;
- vehicle fueling, service, and maintenance facilities;
- vehicle and equipment cleaning facilities;

- fleet storage areas;
- industrial areas;
- marinas - service and maintenance;
- outdoor liquid container storage;
- outdoor loading/unloading facilities;
- public works storage areas;
- facilities that generate or store hazardous materials;
- commercial container nurseries; and
- other land uses and activities as designated by an appropriate review authority

2.1.5 Spill Prevention and Cleanup

This category of BMPs includes programs that reduce the risk of spills during outdoor handling and transportation of chemicals and other materials. It also includes the development of plans and programs to respond, contain and rapidly clean up spills when they do occur so that they do not enter the storm drain system. According to the U.S. EPA (2001), spill response and prevention plans should "clearly state measures to stop the source of a spill, contain the spill, cleanup the spill, dispose of contaminated materials, and train personnel to prevent and control future spills." Such plans are most applicable to construction sites where hazardous wastes are stored or used (U.S. EPA, 1992). The preliminary steps include: (i) identifying potential spill or source areas such as loading and unloading, storage, and processing areas; places that generate dust or particulates; and, areas designated for waste disposal; and, (ii) evaluating stationary facilities that include manufacturing areas, warehouses, service stations, parking lots, and access roads.

2.1.5.1 Above Ground Tank Spill Control

Preventing or reducing the discharge of pollutants to stormwater from storage tanks can be done by installing safeguards against accidental releases, installing secondary containment, conducting regular inspections, and training employees in standard operating procedures and spill cleanup techniques (WEF and ASCE, 1998).

2.1.5.2 Vehicle Spill Control

Methods for preventing or reducing the discharge of pollutants to stormwater from vehicle leaks and spills include reducing the chance for spills by preventive maintenance, stopping the source of spills, containing and cleaning spills, properly disposing of all spill materials, and training employees. The following practices help to reduce the impact from these activities (U.S. EPA, 1999b):

- All spills or leaks should be cleaned using a dry absorbent such as cat litter or commercially available sorbents and disposed of appropriately.
- All used fluids should be recycled or disposed of appropriately.
- All fluid leaks should be repaired as soon as possible to reduce discharge to the environment.

2.1.6 Illegal Dumping Controls

The use of measures to detect, correct, and enforce laws against illegal dumping of pollutants in gutters and streets and into the storm drain system and creeks can have a significant effect on stormwater quality. This BMP includes the control of both direct and indirect sources (WEF and ASCE, 1998). This category of BMPs comprises ordinances, public education programs, and authorized enforcement measures aimed at keeping individuals and businesses from dumping various waste products onto the urban landscape and into the drainage system. The control of illegal dumping practices is important to prevent contaminated runoff from entering wells and surface water, as well as averting flooding due to blockages of drainage channels for runoff (U.S. EPA, 2001). The key to successfully using this BMP is increasing public awareness of the problem and its implications. Some of the issues that need to be examined when creating a program include:

- the locations of persistent illegal dumping activity;
- types of wastes dumped and the profile of dumpers;
- possible driving forces behind illegal dumping, such as excessive user fees, restrictive curbside trash pickup, or ineffective recycling programs;
- previous education and cleanup efforts;
- current control programs and local laws or ordinances addressing the problem; and
- sources of funding and additional resources that may be required.

Illegal dumping controls should focus on the following program areas (U.S. EPA, 1998a): (i) cleanup efforts; (ii) community outreach and involvement; (iii) targeted enforcement; and, (iv) tracking and evaluation. A few well known examples of illegal dumping controls are discussed below:

2.1.6.1 Storm Drain Stenciling

Storm drains frequently discharge runoff directly to water bodies with little or no treatment. Dumping of waste materials into these systems (inlets, catchbasins, channels and creeks) could have severe impacts on receiving water quality. Storm drain stenciling programs that educate residents not to dump materials into storm drains or onto sidewalks, streets, parking lots and gutters is an effective means of reducing nonpoint source pollution associated with such illegal dumping. Storm drain signs and stencils use prohibitive language and/or graphic icons discouraging illegal dumping of improper materials into the urban stormwater runoff conveyance system, and are typically placed directly adjacent to storm drain inlets (WEF and ASCE, 1998; U.S. EPA, 1999b; CASQA, 2003). The use of these highly visible source controls is a common educational strategy in pollution prevention in the U.S. and Australia. Although a few studies reported a positive correlation between seeing the stencils and levels of stormwater awareness, there is no conclusive observation to demonstrate that stormwater drain stenciling raises public awareness or induces behavioral change (Taylor and Wong, 2002).

2.1.6.2 Household Hazardous Waste Collection

Collection programs to recover substances that may otherwise be dumped onto land or into stormwater is a common BMP in major areas. The U.S. EPA (1993a) reported rapid growth in the number of household hazardous chemical collection programs in the U.S., from a mere 2 programs in 1980 to 822 in 1990. Collection efforts should be integrated with an already established municipal solid waste program. The frequency of collection is based on waste type, community characteristics, existing programs and budgets.

2.1.6.3 Used Oil Recycling

Used motor oil is a hazardous waste as it contains heavy metals picked up from the engine during use and should be disposed of at a local recycling or disposal facility. The recycling of used motor oil is a responsible alternative to improper disposal practices such as dumping oil in the sanitary sewer or storm drain system, applying oil to roads for dust control, placing used oil and filters in the trash for landfill disposal, or simply pouring used oil on the ground. Used oil can be recycled in a number of ways (U.S. EPA, 2001):

- Reprocessing used oil into fuel, which is used for heating and cooling homes. This is the most common method of recycling used oil in the U.S. with approximately 750 million gallons being reprocessed every year and marketed to asphalt plants, steel mills, boilers, pulp and paper mills, cement/lime kilns, and a number of other places.
- Motor oil can also be burned in furnaces for heat or in power plants to generate electricity for homes, businesses, or schools.
- It can also be blended for marine fuels, mixed with asphalts for paving, or used in industrial burners.
- Used motor oil can also be used in specially designed municipal garages, space heaters, and automotive bays.
- Used motor oil can be re-refined into lubricating oils that meet the same standards as virgin/new oil.
- Implementation of recycling programs involving public participation at local and national levels

2.1.7 Illicit Connection Controls

Illicit connections are defined as "illegal and/or improper connections to storm drainage systems and receiving waters" (CWP, 1998). The interest in illicit or inappropriate connections to storm drainage systems is an outgrowth of investigations into the larger problem of determining the role of urban stormwater runoff as a contributor to receiving water quality problems. Identifying illicit and improper connections is necessary for all sewer systems, especially in areas where pollutants with unknown sources have been detected in receiving waters. Non-stormwater outfall discharges fall into three categories: (i) pathogenic/toxicant; (ii) nuisance and aquatic life threatening; and, (iii) clean water. Outfall discharges containing pathogenic or toxic pollutants are the most important, and the most likely sources of these pollutants are sanitary and industrial wastewaters (U.S. EPA, 1993c). Potential sources of contaminated entries into storm drainage systems include residential and commercial sources; industrial sources; intermittent sources; direct connections to storm drains; and infiltration to storm drains. A major source of illicit discharges to storm drain systems are direct connections of sanitary sewer piping to the storm drain system. Besides direct connections, seepage and sewage from leaking sanitary sewer lines find their way into storm drains, especially in areas where storm drains run parallel to the sanitary sewer lines. The level and types of industrial activities and the surrounding land uses and ordinances will affect the methods used to identify illicit connections.

2.1.7.1 Illicit Connection - Prevention, Detection and Removal

This group of controls is directed at preventing by ordinance, and eliminating by discovery or removal, connections to the storm drainage system that discharge any material other than stormwater runoff. Bans on connection of floor drains, wash down areas, septic tank overflows, and other similar practices to the stormwater conveyance system are all a part of this BMP category. The prevention of unwarranted physical connections to the storm drain system from sanitary sewers and floor drains through regulation, regular inspection, testing and education can remove a significant source of stormwater pollution. Control procedures for detecting and removing illegal connections from the storm drain conveyance system should be implemented to identify, repair, and remediate infiltration, inflow, and wet-weather overflows from a sanitary sewer to the storm drain conveyance system. This will effectively reduce or prevent unauthorized discharges to receiving waters; some strategies include field-screening, follow-up testing and complaint investigation (U.S. EPA, 1993c, 2003).

Several methods exist for the detection and elimination of illicit cross-connections, which significantly reduce the concentrations of bacteria, nutrients, and oxygen demanding substances contained in stormwater discharges. Useful indicators of the presence of cross connections include dry-weather flows in storm sewer lines and biological indicators that indicate the presence of human fecal matter in storm drain outfalls (U.S. EPA, 1999b). Excavation and correction of illicit connections must be a natural follow-up to detecting these in the first place, and additionally, plans for new development need to be reviewed with periodic inspections during construction preventing future cross connections from being placed.

2.1.7.2 Failing Septic Systems and Sanitary Sewer Overflows

A failing septic system is considered to be one that discharges effluent with pollutant concentrations exceeding established water quality standards. Failure rates for septic systems typically range between 1 and 5 % each year but can be much higher in some regions (U.S. EPA, 2001). Septic system failure has several causes: (i) unsuitable soil conditions; (ii) improper design and installation; or (iii) inadequate maintenance practices. Improperly functioning septic systems contribute significant loads of pollutants (especially nitrogen) and microbial pathogens. Identifying and eliminating these control the contamination of ground and surface water supplies from untreated wastewater discharges.

Sanitary sewer overflows (SSOs) occur in urbanized areas where a separate sanitary sewer system has been created to move wastewater from households and businesses to treatment plants. The detection and elimination of SSOs is most important because sanitary sewer collection systems represent a significant investment for urban municipalities. SSOs can often be reduced or eliminated by a number of practices, including the following (U.S. EPA, 2001):

- sewer system cleaning and maintenance,
- reducing infiltration and inflow through rehabilitation and repair of broken or leaking sewer lines,
- enlarging or upgrading the capacity of sewer lines, pump stations, or sewage treatment plants,
- constructing wet-weather storage and treatment facilities to treat excess flows, and
- addressing SSOs during sewer system master planning and facilities planning.

2.1.8 Stormwater Reuse

Stormwater runoff stored in a surface pond or in the surficial aquifer and then used as a source of irrigation water can reduce potable water use in an area and reduce pollutant loadings from stormwater. The reuse of stormwater is an alternative source of non-potable water and most likely the major use is for irrigation. Stormwater reuse can take place at the (i) household; (ii) municipal; or, (iii) a larger (regional)level (UNEP IETC, 2000).

At the household level, roof runoff could be collected in a tank for use as drinking water (common in arid regions), flushing toilets, or irrigation of gardens. The first flush, generally contaminated with dust particles, leaf litter and animal droppings, can be diverted using a simple diverter; gross particles should be filtered by placing a screen near the inlet. Directing water from the roof directly to garden bed rather than through soakways would benefit shallow-rooted vegetation, especially in arid regions.

At the municipal level, stormwater can be stored in ponds for use for irrigation of parks and gardens and for fire-fighting purposes. Surface detention ponds have been modified to store stormwater for later re-use in Florida (Bradner and Wanielista, 1992; Wanielista and Bradner, 1992). This is in addition to employing the ponds for flood control and for improving the amenity value of the water. Other uses could be for groundwater recharge, water storage during the rainy season and subsequent withdrawal in the dry season. Groundwater recharge can also be used to prevent seawater intrusion in coastal areas subject to heavy groundwater withdrawal in excess of natural replenishment by precipitation.

2.2 Structural or Treatment Best Management Practices

Structural BMPs are used to treat the stormwater either at the point of generation or the point of discharge to either the storm sewer system or to receiving waters. The selection and successful design of selected structural BMPs for stormwater quality enhancement is the cornerstone of stormwater management in newly developing and redeveloping urban areas. Structural BMPs require commitment of resources for initial construction and continuing operation and maintenance. Structural BMPs can be grouped into several general categories; however, the distinction between BMP types and the terminology used to group structural BMPs is an area that requires standardization. For purposes of this paper, and to be consistent with the definitions and terminology used in the ASCE National Stormwater Database and the OW menu of BMPs, structural BMPs for urban stormwater management have been grouped and defined as shown in Table 2-2. (Schueler, 1987; WEF and ASCE, 1998; U.S. EPA, 1999b; NYSDEC, 2001).

Table 2-2. Structural or Treatment Best Management Practices for Urban Stormwater

Major Categories	Structural BMPs
2.2.1 Ponds	▸ Dry Detention Ponds ▸ Dry-Extended Detention Ponds ▸ Wet (Retention) Ponds
2.2.2 Stormwater Wetlands	▸ Constructed Wetlands
2.2.3 Vegetative Biofilters	▸ Grass Swales (Wet/Dry) ▸ Filter Strip/Buffer ▸ Bioretention Cells
2.2.4 Infiltration Practices	▸ Infiltration Trench ▸ Infiltration Basin ▸ Porous Pavement
2.2.5 Sand and Organic Filters	▸ Surface Sand Filter ▸ Perimeter Filter ▸ Media Filter ▸ Underground Filter
2.2.6 Technology Options and Others	▸ Water Quality Inlets ▸ Multi-Chambered Treatment Train ▸ Vortex Separation/Continuous Deflection Systems

(WEF and ASCE, 1998; U.S. EPA, 2001a)

2.2.1 Stormwater Ponds

Stormwater ponds refer to practices that have either a permanent pool of water or a combination of permanent pool and extended detention capable of treating the water quality volume (WQ_v), which is the storage needed to capture and treat the runoff from 90% of average annual rainfall. Treatment of the WQ_v shall be provided at all developments where stormwater management is required.; a more detailed of the same follows in Chapter 3. Stormwater ponds can either be a detention system (dry/extended detention ponds) or a retention system (wet ponds). Detention systems capture a volume of runoff and temporarily retain that volume for subsequent release and do not retain a significant permanent pool of water between runoff events. Retention systems, on the other hand, capture a volume of runoff and retain that volume until it is displaced in part or in total by the next runoff event, thereby maintaining a significant permanent pool of water between runoff events (NYSDEC, 2001).

2.2.1.1 Dry Detention Ponds

Dry ponds, also called detention ponds, are stormwater basins designed to intercept a volume of stormwater runoff and temporarily impound the water for gradual release to the receiving stream or storm sewer system. They are usually designed to reduce the peak flow resulting from a selected design storm (e.g., a 10-yr storm) to the pre-development level to prevent downstream flooding. These typically on-line, end-of-pipe BMPs are designed to completely empty out between runoff events, and therefore mainly provide runoff control as opposed to water quality control (U.S. EPA, 1999b; Metro Council, 2001). Dry ponds are not very effective in removing pollutants, especially particulates, due to the short detention times; particulates that settle to the bottom of the pond are easily re-suspended by subsequent runoff. Dry ponds should therefore be treated as practices used to reduce the peak discharges of stormwater to receiving streams, limit downstream flooding and provide some degree of channel protection.

Most dry ponds are designed to empty in a time period of less than 24 h and can limit downstream scour and loss of aquatic habitat by reducing the peak flow rate and energy of stormwater discharges to the receiving stream. Typically, dry ponds are designed so that release rates are comparable to pre-development flow rates, and their use is largely confined to retrofit situations and as a part of an overall treatment-train approach (Clar *et al.*, 2003).

2.2.1.2 Dry, Extended-detention (ED) Ponds

The outlet structure of a dry pond can be modified in such a way to provide a "retention outlet" that is sized for slow release of the runoff from a designated "BMP storm"; a BMP storm is a small and frequent storm, such as the 1-yr storm, which is prescribed by regulations or ordinances as the BMP design storm (Yu and Nawang, 1993). ED ponds temporarily detain a portion of urban runoff for up to 48 h (a 24 h limit is more common) after a storm, using a fixed orifice to regulate outflow at a specified rate, allowing solids and associated pollutants the required time to settle out. The ED ponds are normally "dry" between storm events, do not have any permanent standing water and typically are composed of two stages: an upper stage, which remains dry except for larger storms, and a lower stage, which is designed for typical storms. The performance can be enhanced by using plunge pools near the inlet, a micropool at the outlet, and an adjustable reverse-sloped pipe as the ED control device (orifice) (NVPDC, 1979; U.S. EPA, 1993a). Temporary and most permanent ED ponds use a riser with an antivortex trash rack on top to control trash.

Advantages:
Dry Ponds:
- Can perform well in cold climates.
- Limit downstream scour and loss of aquatic habitat by reducing the peak flow rate and energy of stormwater discharges to receiving streams.
- Properly designed ponds could be used as recreational areas when not in frequent use.

ED Ponds:
- ED ponds are some of the best facilities for treating spring and winter runoff, as compared

to dry ponds; ponds without ED may have minimal storage above the ice surface; therefore, treatment could be bypassed.

- ED ponds are very effective in controlling peak discharges, which is an important factor in reducing downstream streambank erosion and sediment loads.
- These BMPs are good retrofitting options for existing dry basins and control both stormwater quality and quantity.
- ED ponds that include a dead storage pool can remove significant levels of sediment and sorbed pollutants, and could provide excellent streambank erosion protection and stormwater treatment when used in combination with other structural practices or when retrofitted with permanent pools.

Limitations:
Dry Ponds:

- Dry ponds are not a suitable option for drainage areas of less than 10 acres.
- Dry ponds have a high potential for clogging outlets and sediment resuspension between storm events if improperly maintained, and provide only marginal removal of stormwater pollutants.

ED Ponds:

- Erosion and resuspension of sediments may occur in the pond if the upper stage is not properly vegetated.
- Discharges from ponds may consist of warm water, hence, their use must be limited in areas where warm water discharges will adversely affect a cold-water fishery.

2.2.1.3 Wet (Retention) Ponds

Wet ponds, also known as retention ponds, are designed to intercept a volume of stormwater runoff and provide storage and treatment of this runoff volume. Water in the pond above the permanent pool level, is the volume available for storage (WQ_v), that is displaced in part or completely by the runoff volume from subsequent runoff events. Wet ponds have a capacity greater than the permanent pond volume, which permits storage of the influent stormwater runoff and controlled release of the mixed influent and permanent pond water (Field and Sullivan, 2003). Properly designed and maintained wet ponds can be extremely effective BMPs, providing both water quality improvements and quantity control, in addition to providing aesthetic value and aquatic and terrestrial habitat for a variety of plants and animals (U.S. EPA, 1999a). Wet ponds allow particulate pollutants to settle out and dissolved pollutants to be removed by biological uptake or other decay processes (Yu and Nawang, 1993).

Advantages:

- Properly designed, constructed and maintained wet ponds can provide substantial aesthetic/recreational value and wildlife and wetlands habitat.
- The presence of a permanent wet pool helps provide significant water quality improvement across a relatively broad spectrum of constituents, including dissolved nutrients.
- Widespread application with sufficient capture volume can provide significant control of

channel erosion and enlargement caused by changes to flow frequency relationships that result from the increase of impervious cover in a watershed.

Limitations:

- Wet ponds require relatively large land area and are not suited for drainage areas smaller than 10 acres.
- They cannot be placed on steep or unstable slopes and require a base flow or supplemental flow in order to maintain the water level.
- Improper design and irregular maintenance may result in stratification and anoxic conditions that can promote the release of metals and nutrients from the trapped sediments (CASQA, 2003).

2.2.2. Stormwater Wetlands

Wetlands have long been used in the final treatment of municipal wastewater, and in the last decade have been increasingly used as a stormwater BMP option (Field and Sullivan, 2003). Wetlands remove pollutants through sedimentation, plant uptake, microbial decomposition, sorption, filtration, and exchange capacity, and can be natural, modified natural, or constructed (Metro Council, 2001).

2.2.2.1 Constructed Wetlands

Constructed wetlands are engineered systems designed to simulate the water quality improvement functions of natural wetlands to treat and contain surface water runoff pollutants and decrease loadings to surface waters. Constructed wetland systems are similar to detention and retention systems, except that a major portion of the BMP water surface area (in pond systems) or bottom (in meadow-type systems) contains wetland vegetation and this system also includes wetland channels (NYSDEC, 2001).

Constructed wetland systems incorporate the natural functions of wetlands to aid in pollutant removal from stormwater and also control stormwater quantity by providing a significant volume of ponded water above the permanent pool elevation. However, constructed wetlands treating urban runoff differ from artificial wetlands created to comply with mitigation requirements in that they do not replicate all of the ecological functions of natural wetlands. Enhanced designs may include a forebay, complex microtopography, and pondscaping with multiple species of wetland trees, shrubs, and plants (CASQA, 2003). Incorporating a sediment forebay or some other pretreatment provision into the wetland design allows for the removal of coarse sediments, which may otherwise degrade the performance of the system.

The use of stormwater wetlands is limited by a number of site constraints, including soil types, depth to groundwater, contributing drainage area and available land area. Constructed wetlands are especially appropriate where groundwater levels are close to the surface as groundwater can supply the necessary water required for sustaining the wetland system. It has been observed that medium-fine texture soils such as loams and silt loams are best to establish vegetation, retain surface water, permit groundwater discharge and capture pollutants (Metro Council, 2001).

Advantages:

- Flood attenuation, reduction of peak discharges and improvements in downstream water quality.
- Settlement of particulate pollutants, reduction of oxygen-demanding substances and bacteria from urban runoff, and biological uptake of pollutants by wetland plant species.
- Enhancement of vegetation diversity and wildlife habitat in urban areas as well as aesthetic enhancement and valuable addition to community green space.

Limitations:

- Maintenance of wetland vegetation may be difficult under a variety of flow conditions.
- Pollutant removal efficiencies may be lower than anticipated until the establishment of wetland vegetation.
- The release of nutrients in the fall could have an impact on receiving water quality.
- Wetlands may act as a heat sink and discharge warmer water to downstream water bodies.

2.2.3 Vegetated Systems (Biofilters)

Vegetated systems are practices explicitly designed to capture and treat the full WQ_v within dry or wet cells formed by check dams or other means and include systems designed to convey and treat either shallow flow (swales) or sheetflow (filter strips) runoff. These BMPs are commonly referred to as "biofilters," since the grasses and vegetation "filter" the stormwater as it flows over them (U.S. EPA, 1999b). A certain degree of treatment, storage and infiltration could be provided by conveying stormwater runoff in vegetated systems, which helps to reduce the overall volume of stormwater runoff generated from a particular drainage area; this makes open channel vegetated systems a better alternative to traditional curb-and-gutter and storm sewer conveyance systems (CASQA, 2003). While biofiltration swales are vegetated channels that receive directed flow and convey storm water, biofiltration strips, also known as vegetated buffer strips, are vegetated sections of land over which stormwater flows as overland sheet flow (Caltrans, 2002). Pollutants are removed by filtration through the grass, sedimentation, adsorption to soil particles, and infiltration through the soil (Field and Sullivan, 2003). Strips and swales are mainly effective at removing debris and solid particles, although some dissolved constituents are removed by adsorption onto the soil (Metro Council, 2001).

2.2.3.1 Grass Swales (Wet, Dry)

A grassed swale is an infiltration/filtration method that is usually used to provide pretreatment before stormwater runoff is discharged to treatment systems. Grassed swales are typically shallow, vegetated, man-made conveyance channels designed so that the bottom elevation is above the water table to allow runoff to infiltrate into ground water. The vegetation or turf covering the side slopes and channel bottom collect and slowly convey runoff to downstream discharge points (CASQA, 2003). Swales are designed to treat stormwater runoff through filtering by the vegetation in the channel, filtering through a subsoil matrix, and/or infiltration into the underlying soils, during which they trap pollutants, promote infiltration, and reduce flow velocities. Swales can be either wet or dry; while dry swales are used in areas where standing water is not desired, such as residential areas, wet swales can be used where standing water does not create a nuisance problem and where the groundwater is close enough to the surface to maintain the permanent pool in inter-event periods. The added benefit of wet swales is the ability to include in them a range of wetland vegetation to aid in pollutant removal (U.S. EPA, 1999b).

2.2.3.1.1 Dry Swales

The dry swale is a type of open vegetated channel used to treat and attenuate the WQ_v of stormwater runoff, as well as convey excess stormwater downstream. The entire WQ_v of a given storm is temporarily held in a pool or series of pools created by permanent checkdams or ditchblocks. This holding time serves to settle pollutants, especially sediments. Typically located in a drainage easement at the back side of a residential lot or along roadsides in place of curbs and gutters, dry swales are good options in residential settings, as they discourage long-standing water. This makes it possible to mow the area shortly after a rainfall event. Stormwater treated by the soil bed flows into an underdrain system that conveys treated stormwater back to the storm drain system (Metro Council, 2001).

Advantages:

- Good option for small-area stormwater retrofits, replacing existing drainage ditches; for residential or institutional areas of low to moderate density, and may also be used in parking lots to break up impervious areas.
- Linear nature of the design works well for treating highway or residential road runoff.
- Rapid de-watering and shallow slopes that are easy to mow.
- Since runoff ponds for only a short time in a dry swale, water temperatures do not significantly increase, making this an appropriate practice for use in watersheds with cold-water trout streams.

Limitations:

- Individual dry swales can treat only a small area and may not be applicable to sites with many driveway culverts or extensive sidewalk systems; roadside dry swales are subject to damage from off-street parking and snow removal (Metro Council, 2001).
- Do not appear to be effective in reducing bacteria levels in stormwater, and appear to remove only modest amounts of phosphorus, with relatively few studies known to gauge their effectiveness.

2.2.3.1.2 Wet Swale

The wet swale, also called a grassed open channel, consists of a broad open channel capable of temporarily storing water, and unlike the dry swale, does not have an underlying filtering bed. The wet swale is constructed directly within existing soils and may or may not intersect the water table. Water quality treatment mechanisms rely primarily on settling of suspended solids, adsorption, and microbial breakdown of pollutants. Like the dry swale, the entire water quality treatment volume is stored and retained within a series of cells in the channel, formed by berms or checkdams. Wet swales also reduce the velocity of stormwater runoff and may promote infiltration (Metro Council, 2001). A wet swale is a suitable option when the water table is located very close to the surface, with the result that swale soils often become fully saturated, or have standing water all or part of the year after the channel has been excavated. Thus a wet swale essentially acts as a very long and linear shallow wetland treatment system. Pollutant removal rates in some cases may be enhanced by planting emergent wetland plant species in these cells (CASQA, 2003).

Advantages:

- Control peak discharges by reducing runoff velocity and promoting infiltration and provide effective pretreatment for BMPs in series by trapping, filtering and infiltrating pollutants.
- Convey water in properly protected channels and divert water around potential pollutant sources.
- Provide water quality treatment by sedimentation and biological uptake and enhance biological diversity and create beneficial habitat between upland and surface waters.

Limitations:

- Impractical in areas with very flat grades, steep topography, wet or poorly drained soils.
- Possibility of erosion when flow volumes and/or velocities are high during storm events.
- Area requirements can be excessive for highly developed sites.
- Roadside wet swales become less feasible with an increase in the number of driveway entrances requiring culverts.

2.2.3.2 *Vegetative Filter Strips (VFS)*

According to the U.S. EPA (1993a), a vegetative buffer area or strip is defined in many cases as a "permanent, maintained strip of planted or indigenous vegetation located between nonpoint sources of pollution and receiving water bodies for the purpose of removing or mitigating the effects of diffuse source pollutants such as nutrients, pesticides, sediments, and suspended solids." These are vegetated strips of land that act as "buffers" by accepting storm runoff as overland sheet flow from upstream developments and providing similar treatment potential mechanisms to that of swales, prior to discharge of the storm runoff to the storm drainage system (Field and Sullivan, 2003). They may closely resemble many natural ecotones, such as grassy meadows or riparian forests, and their dense vegetative cover facilitates sediment attenuation and pollutant removal. Vegetative filter strips (VFS) are frequently planted with turf grass; however, alternatives that adopt any natural vegetated form such as meadows or small forest may be used. Originally used as an agricultural practice, filter strips are now evolving as an urban practice (CASQA, 2003). However, VFS, unlike grassed swales, are effective only for overland sheet flow and provide little treatment for concentrated flows. Therefore, they are more suitable for use in agriculture and low-density development, as well as other situations where runoff tends not to be concentrated. VFS are often used as pretreatment for other structural practices, such as infiltration basins and infiltration trenches, and land grading and/or a level spreader can be used to create a uniformly sloping area that distributes the runoff evenly across the filter strip (U.S. EPA, 1993a). They can also be used in combination with riparian buffers in treating sheet flows and in stabilizing drainage channel banks and stream banks. In semi-arid climates, grass filter strips may need to be irrigated to maintain a dense stand of vegetation and to prevent export of unstabilized soil (Metro Council, 2001).

Advantages:
- VFS work well in residential areas, where they provide open space for recreation activities, help maintain riparian zones along streams, reduce streambank erosion and provide animal habitat.
- VFS can be useful as sediment filters during construction, which in some cases, may require only the preservation of an appropriately located area of existing vegetation.
- Flow characteristics and vegetation type and density can be closely controlled to maximize BMP effectiveness.

Limitations:
- VFS cannot treat a very large drainage area and they require a thick vegetative cover to ensure proper functioning.
- This BMP does not provide significant attenuation of the increased volume and flow rate of runoff during intense rain events and is best implemented as one of a series of stormwater BMPs.
- Vegetative buffers may not provide treatment for dissolved constituents except to the extent that flows across the vegetated surface are infiltrated into the soil profile.
- VFS is not recommended to treat highly contaminated "hotspot" runoff, since infiltration could result in groundwater pollution and damage to vegetation.

2.2.3.3 Bioretention

Bioretention systems are designed to mimic the functions of a natural forest ecosystem for treating stormwater runoff. According to one definition, "bioretention systems are a variation of a surface sand filter, where the sand filtration media is replaced with a planted soil bed" (U.S. EPA, 1999b). Another approach considers bioretention as a "concept that uses biologic activity (plants and microbes) to filter/clean stormwater by being incorporated into different kinds of infiltration and filtration BMP designs such as infiltration basins, rainwater gardens and surface sand filters" (Metro Council, 2001). This book views bioretention as a BMP that functions as a soil and plant-based filtration device for stormwater management and removes pollutants through a variety of physical, biological and chemical treatment processes. These facilities normally consist of a grass buffer strip, sand bed, ponding area, organic layer or mulch layer, planting soil and plants.

In general, bioretention systems can be described as shallow, landscaped depressions commonly located in parking lot islands or within small pockets in residential areas that receive stormwater runoff. Stormwater flows into the bioretention area, ponds on the surface, and gradually infiltrates into the soil bed. A number of processes including adsorption, filtration, volatilization, ion exchange and decomposition are responsible for pollutant removal (PGDER, 1993). Filtered runoff can either be allowed to infiltrate into the surrounding soil (functioning as an infiltration basin or rainwater garden), or collected by an underdrain system and discharged into the storm sewer system or directly to receiving waters (functioning like a surface sand filter). Runoff from larger storms is generally diverted past the area to the storm drain system. The bioretention system has been used as a stormwater BMP since 1992. In addition to Prince George's County, MD and Alexandria, VA, bioretention has been used successfully in urban and suburban areas in Montgomery County, MD;

Baltimore County, MD; Chesterfield County, VA; Prince William County, VA; Smith Mountain Lake State Park, VA; and Cary, NC (CASQA, 2003).

A bioretention system includes the following components (Metro Council, 2001):
1. Grass Buffer Strips: Runoff enters the bioretention area as sheet flow through the grass buffer strips where the buffers reduce runoff velocity and filter particulates from the runoff.

2. Ponding Area: The ponding area provides for surface storage of storm runoff before it filters through the soil bed. It also allows for the evaporation of ponded water as well as the settling of sediments in the runoff.

3. Organic Mulch Layer: The organic mulch layer protects the soil bed from erosion, retains moisture in the plant root zone, provides a medium for biological growth and decomposition of organic matter and provides some filtration of pollutants.

4. Planting Soil Bed: The planting soil bed provides water and nutrients to support plant life in the bioretention system. Stormwater filters through the planting soil bed where pollutants are removed by filtration, plant uptake, adsorption and biological degradation.

5. Sand Bed: The sand bed underlies the planting soil bed and allows water to drain from the planting soil bed through the sand bed and into the surrounding soil. It also provides additional filtration and allows for aeration of the planting soil bed.

6. Plants: Plants are important components of the bioretention system, and remove water through evapotranspiration and pollutants and nutrients through uptake. The plant species selected are designed to replicate a forested ecosystem and to survive stresses such as frequent periods of inundation during runoff events and drying during inter-event periods.

Advantages:
- Properly designed and maintained bioretention systems are more likely to be aesthetically pleasing than other types of filtration or infiltration systems due to the presence of plants.
- Layout of bioretention facilities can be very flexible, and the selection of plant species can provide for a wide variety of landscape designs.
- Ideally suited to highly impervious areas, such as parking lots, and can be applied in many different climates and geologic environments, with some minor design modifications.
- Reduce the volume of runoff from a drainage area and can be very effective for removing fine sediments, trace metals, nutrients, bacteria and organics (U.S. EPA, 1999a).

Limitations:
- Bioretention as a BMP is not recommended for areas with slopes greater than 20% or where mature tree removal would be required since clogging may result, particularly if the facility receives runoff with high sediment loads.
- Bioretention is not a suitable BMP in locations where the water table is within 6 feet of the ground surface and where the surrounding soil stratum is unstable.

- Susceptible to clogging by sediment; therefore pretreatment is a necessary part of design. In cold climates, the soil may freeze, preventing runoff from infiltrating into the planting soil.
- By design, bioretention BMPs have the potential to create very attractive habitats for mosquitoes and other vectors because of their highly organic, often heavily vegetated areas mixed with shallow water.

2.2.4 Infiltration Systems

Infiltration practices have a high potential of controlling stormwater runoff by disposal at a local site level. An infiltration BMP is designed to capture a volume of runoff, retain it and allow it to infiltrate into the ground. There are a number of advantages and disadvantages associated with infiltration practices (Schueler, 1987). Advantages include the control of both the quality and quantity of stormwater. Water quantity is controlled by controlling surface runoff infiltration into the underlying soil. This reduces the volume of water discharged to receiving streams and thereby some of the potential impacts caused by excess flows, as well as reducing increased pollutant concentrations in the receiving stream. They can be designed to capture a volume of stormwater and infiltrate this volume into the ground over a period of several hours or even days, thereby maximizing the infiltrating capacity of the BMP. Secondary benefits include increased recharge of underlying aquifers, increasing base flow levels of nearby streams, and water quality treatment. Pollutant removal occurs as water percolates through the various soil layers and particulates are filtered out in addition to soil microbial degradation of organic pollutants contained in the infiltrated stormwater. The disadvantages are just as numerous. Infiltration may not be appropriate in areas where groundwater is a primary source of drinking water due to this method's potential for contaminant migration. This holds true when runoff is from a commercial or residential area with a higher potential for metal or organic contamination. Also, the performance is limited in areas with poorly permeable soils, and these BMPs can experience reduced infiltrating capacity and clogging due to excessive sediment accumulation.

2.2.4.1 Infiltration Basins

Infiltration basins are typically off-line, end-of-pipe BMPs, and are designed to intercept only a certain volume of runoff. A flow splitter or weir is usually used to divert runoff from a storm sewer system into the infiltration basin. The basin may or may not be lined with plants. However, vegetated infiltration systems help to prevent migration of pollutants, and the roots of the vegetation can increase the permeability of the soils, thereby increasing the efficiency of the basin. The main purpose of this BMP is to transform a surface water flow into a groundwater flow and to remove pollutants through mechanisms such as filtration, adsorption and biological conversion as the water percolates through the underlying soils. Infiltration basins should drain within 72 h to maintain aerobic conditions that favor bacteria that aid in pollutant removal, and to ensure that the basin is ready to receive the next storm (Schueler, 1987; U.S. EPA, 1993b).

Infiltration basins are inappropriate for areas that contribute high concentrations of sediment or suspended solids without adequate treatment. Pretreatment, such as grit chambers, swales with check dams, filter strips, or a sedimentation basin should be a fundamental component of any BMP system relying on infiltration. Runoff entering the basin is pretreated to remove coarse sediment that may clog the surface soil pores on the basin floor. Concentrated runoff should flow through a sediment trap, or a filter strip may be used for sheet flow.

Infiltration basins are dry ponds constructed to allow infiltration to occur simultaneously with other treatment processes. The operating characteristics of infiltration basins are essentially the same as for dry detention ponds (Field and Sullivan, 2003), except a few significant exceptions (Clar *et al.*, 2003):

- Infiltration basins also remove dissolved and colloidal solids in the volume of infiltrated water, whereas dry ED ponds can remove only the fraction of colloidal solids sorbed to settleable solids.
- The settling velocities of sediment particles and particulate (settleable) chemicals are increased by a value equal to the infiltration rate in the basin. The impact would be more for clay (colloidal) sized particles than for silt, sand, and small or large aggregates.
- Infiltration practices differ from typical dry basins because they contribute to groundwater recharge, thereby providing an additional element of performance.
- By providing volume control, these BMPs can effectively address the increased frequency and duration of peak flows and can provide downstream channel protection.
- The mode of operation being infiltration of runoff into subsurface soils, infiltration basins are able to preclude the thermal impacts issues associated with detention, ED and wet ponds.

Advantages:

- The principal benefit of infiltration basins is the approximation of pre-development hydrology during which a significant portion of the average annual rainfall runoff infiltrates and evaporates rather than flushing directly to creeks.
- Infiltration basins can be useful for controlling channel forming (erosion) and high frequency (generally less than the 2-yr) flood events by adequately sizing WQ_v.
- Provides groundwater recharge and baseflow in nearby streams, reduces local and downstream flooding and protects streambank integrity.
- Can be very effective for removing fine sediment, trace metals, nutrients, bacteria and oxygen-demanding substances.

Limitations:

- Potentially high failure rates due to improper siting, design and lack of maintenance, especially if pretreatment is not incorporated into the design.
- Not appropriate for treating significant loads of sediment and other pollutants due to the potential for clogging.
- Not suitable on fill sites or steep slopes and a risk of groundwater contamination in very coarse soils.

2.2.4.2 Infiltration Trenches

Infiltration trenches are shallow (3 to 12 ft) excavated ditches that are lined with filter fabric and filled with stone to create underground reservoirs for stormwater runoff from a specific design storm. Urban runoff diverted into the trench gradually infiltrates through the bottom and sides of the trench into the subsoil and eventually into the groundwater over a period of days (Metro Council, 2001). Infiltration trenches are typically implemented at the ground surface to intercept overland flows. Runoff can be captured by depressing the trench surface or by placing a berm at the down gradient side of the trench (Schueler, 1987). Design variations include dry wells, pits designed to control small volumes of runoff (such as the runoff from a rooftop), and enhanced infiltration trenches, which are equipped with extensive pretreatment systems to remove sediment and oil. Depending on the quality of the runoff, pretreatment will generally be necessary to lower the failure rate of the trench. More expensive than pond systems in terms of cost per unit of runoff treated, infiltration trenches are best suited for drainage areas of less than 5 to 10 acres, or where ponds cannot be applied (Schueler, 1992).

The design storm for an infiltration trench is typically a frequent, small storm such as the 1-yr event, that provides treatment for the "first flush" of stormwater runoff. Hence they are frequently used in combination with other BMPs such as a detention basin to control peak hydraulic flows. Infiltration trenches provide total peak discharge, runoff volume and water quality control for all storm events equal to or less than the design storm. This infiltration reduces the runoff volume, removes many pollutants and provides stream baseflow and groundwater recharge (CASQA, 2003).

Advantages:

- Provides 100% reduction in the load discharged to surface waters.
- The principal benefit of infiltration basins is the approximation of pre-development hydrology during which a significant portion of the average annual rainfall runoff is infiltrated and evaporated rather than flushed directly to creeks.
- If the WQ_v is adequately sized, infiltration basins can be useful for providing control of channel forming (erosion) and high frequency (generally less than the 2-yr) flood events.
- As an underground BMP, trenches are unobtrusive and have little impact on site aesthetics.

Limitations:

- High failure rate if soil and subsurface conditions are unsuitable.
- Unsuitable on fill sites or steep slopes and a risk of groundwater contamination in very coarse soils.
- Upstream drainage area must be completely stabilized before construction.

2.2.4.3 Porous Pavement

Porous pavement is an infiltration system in which stormwater runoff infiltrates into the ground through a permeable layer of pavement or other stabilized permeable surface. The permeable surface can include porous asphalt, porous concrete, modular perforated concrete block, cobble pavers with porous joints, gaps or reinforced/stabilized turf (Florida DER, 1988). Concrete grid pavement consists of concrete blocks with regularly inter-dispersed void areas with pervious materials, such as gravel, sand, or grass. The blocks are typically placed on a sand or gravel base and designed to provide a load-bearing surface that is adequate to support vehicles, while allowing infiltration of surface water into the underlying soil. Porous pavement provides an alternative to conventional pavement and reduces much of the need for urban runoff drainage conveyance and treatment offsite. Instead, runoff is diverted through a porous asphalt layer into an underground stone reservoir before gradually exfiltrating out of the stone reservoir into the subsoil (Field and Sullivan, 2003). Porous pavements tend to clog with fine sediments and need to be subjected to periodic vacuum-type street sweeping (WA DOE, 1991). Modular pavers, grassed parking areas, and permeable pavements may also be employed to reduce runoff volumes and trap vehicle-generated pollutants (U.S. EPA, 1993a); however, care should be taken when selecting such alternatives. The potential for groundwater contamination, compaction, or clogging due to sedimentation should be evaluated during the selection process. These practices should be selected only in cases where proper operation and maintenance can be guaranteed due to high failure rates without proper upkeep.

There has been mixed observations on the performance of porous pavement systems. The performance of porous asphalt has been historically very poor in the mid-Atlantic region. However, many of these failures can be attributed to lack of proper erosion and sediment controls during construction or lack of contractor experience with installation of porous pavement systems. Yet, studies on porous concrete systems in use in Florida show that they have performed very well (U.S. EPA, 1999b).

Advantages:

- When properly designed and maintained, porous pavement systems can be an effective means of managing stormwater runoff, and particularly useful for overflow parking areas that are not used on a daily basis.
- Especially useful for driveways and streets and in residential areas, and in commercial parking lots.

Limitations:

- Not effective in areas that receive runoff with high amounts of sediments due to the tendency of the pores to clog.
- Require maintenance, including periodic vacuuming or jet-washing to remove sediment from the pores.

2.2.5 Filtration Systems

A filtration system is a device that uses a combination of a granular filtration media such as sand, gravel, organic material (e.g., peat, compost), membrane, or other acceptable treatment media to remove pollutants in stormwater runoff. Sand filters are a self-contained bed of sand to which the first flush of runoff is diverted. The runoff percolates through the sand, where colloidal and particulate materials are strained out by the cake of solids that forms, or is placed on the surface of the media. Water leaving the filter is collected in underground pipes and returned to the stream or channel. A layer of peat, limestone, and/or topsoil may be added to improve removal efficiency

(CASQA, 2003). Detention time is typically 4 to 6 hs in a filter, and sediment-trapping structures are typically used to prevent premature clogging of the filter media (NVPDC and ESI, 1992). Filtration systems are primarily water quality control BMPs designed to remove particulate pollutants in stormwater; quantity control can also be included by providing additional storage volume in a pond or basin, or vertical storage volume above the filter bed, or by allowing water to temporarily pond in parking lots or other areas before being discharged to the filter. Sand filters have been demonstrated to be effective in removing many of the common pollutants found in urban stormwater runoff, especially the particulate pollutants. While they seem to have a moderate level of bacterial removal, they have not been effective at removing total dissolved solids and nitrate-nitrogen (Metro Council, 2001). Media filters can be successfully employed for stormwater management in (i) small sites such as parking lots and developments; (ii) areas with high pollution potential such as industrial areas; or, (iii) highly urbanized areas, where land availability or costs preclude the use of other BMP types (U.S. EPA, 1999b). Filters should be placed off-line, (i.e., the water quality volume is diverted to the BMP, while any flows in excess of this volume are bypassed) and are sometimes designed to intercept and treat only the first flush and bypass larger storm flows (CASQA, 2003).

Commonly used types of filters include: (i) surface sand filters such as the "Austin" sand filter; and (ii) underground vault filters such as the "DC" sand filter and the "Delaware" sand filter. Design variations of these basic types exist in addition to several proprietary filtering systems; some use specialized filter media made from materials such as leaf compost (Metro Council, 2001). This is in addition to a number of variations in the types of filtration media used described earlier. These designs may also incorporate additional features such as a layer of filter cloth or a plastic screen, gravel layer, peat layer, compost layer, layer of peat or a peat/sand mixture.

2.2.5.1 Surface Sand Filter

The surface sand filter was initially developed in Florida in 1981 for sites that could not infiltrate runoff or were too small for effective use of detention systems, with further development of filter technology by the city of Austin, Texas. A surface sand filter consists of a pretreatment basin, water storage reservoir, flow spreader, sand and underdrain piping. A basin liner may also be needed if the treated runoff is not to be allowed to infiltrate into the soil underlying the filtration basin because of groundwater concerns. This system usually incorporates two basins: The first is a sedimentation basin, dry or wet, where runoff enters and is removed of coarse particulates by gravity settling. Water then flows over a weir or through a riser into the filter basin, which consists of sand with a gravel and perforated pipe underdrain system to capture the treated water. The surface of the filter bed may be planted with grass. Additional storage volume is provided above the filter bed to increase the volume of water that can be temporarily ponded in the system prior to infiltration. This two basin configuration helps to limit premature clogging of the filter bed due to excessive sediment loading. There are several variations of this design, Austin uses two variations, partial, and full sedimentation filtration systems (City of Austin, 1988).

2.2.5.2 Underground Vault Sand Filter

An underground filter is similar to a surface filter except that the sand (or other media) and underdrains are installed below grade in a vault. The underground vault sand filter was developed by the District of Columbia in the 1980s. Its design incorporates three chambers. The first chamber and the throat of the second contain a permanent pool of water and function as a sedimentation chamber and an oil and grease and floatables trap, besides providing temporary runoff storage. The two chambers are connected by a submerged opening or inverted elbow near the bottom of the dividing wall, which provides a water seal that prevents the transfer of oil and floatables to the second chamber containing the filter bed. During a storm event, water flows through the opening into the second chamber and onto the filter bed. Additional runoff storage volume is provided above the filter bed. Filtered water is collected by a gravel and perforated pipe underdrain system and flows into the third chamber, which contains a clearwell and is connected to the storm drain system. Overflow protection can be provided by placing the filter off-line or by providing a weir at the top of the wall connecting the filter chamber with the clearwell chamber to serve as an overflow. The schematic of an underground vault sand filter is presented in Chapter 3 (Figure 3-11).

2.2.5.3 Perimeter Filter

A perimeter filter is an underground vault sand filter, and is referred to as a perimeter sand filter because of its particular suitability for use around the perimeters of parking lots. This system, developed in Delaware, and also known as the "Delaware Sand Filter", contains two chambers and a clear well. Stormwater enters the first chamber, which serves as a sedimentation chamber. Water then flows over a series of weirs and into the second chamber, which contains the filter media. Additional storage volume is provided by water temporarily ponding in both chambers. Filtered water is collected by a series of gravel and perforated pipe underdrains, and flows into a clearwell that contains a connection to the storm drain system.

2.2.5.4 Others

In addition to the three basic filtering systems, in DC, Austin, Delaware, there are a number of variations in use. A compound stormwater filtering system developed by the city of Alexandria, VA incorporates an anoxic filtration zone in a permanently flooded gravel layer in the filter. This anoxic zone aids in nitrogen removal by anoxic denitrification. Another configuration uses an upflow anaerobic filter upstream of the sand filter to enhance phosphorus removal by precipitating more iron on the sand filter. Yet another type employs organic materials such as peat or compost combined with sand or other materials. Filters that use an organic filtration media, such as peat or leaf compost, are useful in areas where additional nutrient or metal control is desirable due to the adsorptive capacity, its non-exchange capability, and its ability to serve as a medium for the growth of a variety of microorganisms. However, peat must be carefully selected (fibric and/or hemic, not sapric, peat should be used) and the environmental concerns regarding destroying peat bogs to obtain filtration media need to be addressed when other technologies are available (Schueler, 1987; VA DCR, 1999).

Advantages:

- Media filters can be used in high density urban sites with small drainage areas that are completely impervious such as parking lots. They can be applicable to many areas that are difficult to retrofit due to space limitations and used in sites where soil or groundwater concerns do no support an infiltration BMP.
- Applicable in small drainage areas of 1 to 10 acres; take up little space and can be used on highly developed and steeply sloped sites.
- Provide high removal efficiencies for total suspended solids (TSS).

Limitations:

- Pretreatment required to prevent the filter media from clogging.
- Not applicable in areas of high water tables and should not be used in areas where heavy sediment loads are expected or in tributary areas that are not fully stabilized.

2.2.6 Technology Options and Others

This group of BMPs includes a variety of proprietary and miscellaneous systems that are used for urban stormwater management and that do not seem to fall exclusively under any of the above-mentioned categories. These include water quality inlets, hydrodynamic devices, filtration devices, etc. Many of these systems are "drop-in" systems, and incorporate some combination of filter media, hydrodynamic sediment removal, oil and grease removal, or screening to remove pollutants from stormwater. There are also quite a number of proprietary devices in this category, but this book does not focus on vendor-supplied systems and other proprietary devices due to the lack of peer reviewed performance data for these systems.

2.2.6.1 Water Quality Inlets

Water Quality Inlets (WQI), also commonly called trapping catchbasins, oil/grit separators or oil/water separators, consist of one or more chambers that promote sedimentation of coarse materials and separation of free oil, as opposed to emulsified or dissolved oil, from stormwater. Some WQIs also contain screens to help retain larger or floating debris, and many of the newer designs also include a coalescing unit that helps promote oil/water separation. A typical WQI consists of a sedimentation chamber, an oil separation chamber and a discharge chamber (U.S. EPA, 1999a; Metro Council, 2001).

WQIs are underground retention systems designed to remove settleable solids, and there are several design variations. The simplest form of design is a single-chambered urban runoff inlet in which the bottom has been lowered to provide 2 to 4 ft of additional space between the outlet pipe and the structure bottom for collection of sediment. Some WQIs include a second chamber with a sand filter to provide additional removal of finer suspended solids by filtration; the first chamber provides effective removal of coarse particles and helps prevent premature clogging of the filter media. Other WQIs include an oil/grit separator. A typical oil/grit separator consists of three chambers: The first chamber removes coarse material and debris; the second chamber provides separation of oil, grease, and gasoline; and the third chamber provides safety relief should blockage occur (NVPDC and ESI, 1992). WQIs typically capture the first portion of runoff for treatment and are generally used for pretreatment before discharging to other BMPs (Schueler *et al.*, 1992).

Advantages:

- WQIs can effectively trap trash, debris, oil and grease, and other floatables that would otherwise be discharged to surface waters.
- A properly designed and maintained WQI can serve as an effective BMP for reducing hydrocarbon contamination and spills in receiving water sediments.
- WQIs are generally recommended for drainage areas of 1 acre or less and are effective for industrial hotspots that have the potential for petroleum-contaminated process washdown, spills, and stormwater runoff.

Limitations:

- WQIs have limited ability to separate dissolved or emulsified oil from runoff and are also not very effective at removing nutrients and heavy metals, except where the metals removal is directly associated with sediment removal.
- High sediment loads can interfere with the ability of the WQI to effectively separate oil and grease from the runoff and during high flow conditions, sediment residuals may be resuspended and released from the WQI to surface waters.
- WQIs generally provide limited hydraulic and residuals storage as a result of which they do not provide substantial stormwater improvement.
- Standing water in the devices can be a breeding ground for mosquitoes and other vectors, and lack of maintenance often results in resuspension of settled pollutants.

2.2.6.2 Multi Chambered Treatment Train (MCTT)

A multiple treatment system uses two or more BMPs in series and the MCTT is an example of a stormwater device that utilizes a combination of processes, especially pretreatment of stormwater using sedimentation, followed by media filtration (Field and Sullivan, 2003). The MCTT was developed during an U.S. EPA-sponsored research program on treatability of stormwater at critical source areas, and areas designated as stormwater hotspots; typical locations include gas stations, junkyards, bus barns, public works yards, car washes, oil change facilities, transmission repair shops, auto repair facilities, fast-food restaurants, convenience stores etc., and other areas where stormwater has a high probability of high concentrations of oils and other toxic organic pollutants that are difficult to treat by other means (Pitt *et al.*, 1999). The MCTT contains aeration, sedimentation, sorption and sand-peat filtration, and is most suitable for use at relatively small and isolated paved critical source areas (0.25 to 2.5 acre), where surface land is not available for stormwater controls (Field and Sullivan, 2003). The MCTT consists of three chambers:

- a catchbasin or grit chamber for the removal of large particles and litter (screening),
- a settling chamber for quiescent settling of fine settleable solids, and
- a sand-peat moss "filter" for final polishing (removal of dissolved pollutants).

The collected runoff is first treated in a catchbasin chamber where larger particles are removed by settling. The water then flows into a main settling chamber containing oil sorbent material where it undergoes a much longer treatment period (24 to 72 h) to remove finer particles and associated pollutants. The final chamber contains a mixed media filter material comprising equal amounts of sand and peat. This chamber acts as a polishing "filter" to remove some of the filterable toxicants from runoff by other processes, such as ion exchange and sorption. The MCTT was designed to remove pollutants of a specific class of concern in stormwater: particulates as small as a few µm and associated particulate bound toxicants, plus filterable toxicants. Pilot and full scale test results showed substantial reductions in both particulate and dissolved phases of stormwater toxicants as well as suspended solids. While substantial reductions are observed in the concentration of pollutants and suspended solids in the main settling chamber as well as the peat-sand chamber, the catchbasin/grit chamber did not provide any significant improvements in water quality, although it is an important element in reducing maintenance problems by trapping bulk material (GVSDD, 1999). Additional information on the performance data of MCTTs is provided in (Pitt *et al.*, 1999; Field and Sullivan, 2003).

Advantages:

- The MCTT is capable of reducing a broad range of stormwater pollutants in particulate fractions (as small as a few µm) and filterable fractions that cause substantial receiving water problems.
- Although unit construction cost appears to be expensive (construction costs estimated at $15,000 to $30,000 per 0.1 ha drainage area plus 11 to 13% of construction costs for operation and maintenance per yr (GVSDD, 1999), with right planning and implementation, the MCTT has a high potential for cost-effective use as an integrated component in watershed management programs designed to protect and enhance receiving waters.
- They can be used in areas where site conditions prevent the use of other BMPs such as infiltration, wet ponds and wetlands and in space-limited areas because of the underground design.
- Relatively few concerns associated with aesthetics and safety.

Limitations:

- The MCTT is designed for contaminant removal only and provides no significant attenuation of peak flows or runoff volumes.
- Lack of maintenance and undersized separators for the flows encountered.
- Scouring of previously captured material is commonly observed.

2.2.6.3 Vortex Separators and Continuous Deflective Systems

Several solid/liquid separators use circular flow patterns to create inertial flow separation. One general classification of these devices includes swirl and vortex separators that have been in use for over 30 years to control combined sewer overflows (CSOs). Vortex separators, or swirl concentrators, are gravity separators, and in principle are essentially wet vaults. The difference between them and wet vaults, however, is that vortex separators are round rather than rectangular, and water moves in them in a centrifugal fashion rather than a straight line before exiting. By this approach, it is possible to obtain significant removal of suspended sediments and attached pollutants with less space (U.S. EPA, 1977; Field *et al.*, 1997). These systems were originally developed to treat CSOs for the removal of coarse inorganic solids. They are increasingly being used to control pollution from stormwater discharge (Konieek *et al.*, 1996; Field *et al.*, 1997; CASQA, 2003) by reducing solids concentration in urban runoff (Lee *et al.*, 2003).

Hydrodynamic vortex separation is a proven technology with an established track record for improving urban water quality (including CSOs, stormwater and wastewater) (Andoh and Saul, 2002). Hydrodynamic vortex separators are characterized by tangential flows into a cylindrical vessel, which in turn creates a complex rotary regime and provides three main functions: flow regulation, settleable solids concentration and the capture of floatables (U.S. EPA, 1977; Field *et al.*, 1997). Stormwater enters the vortex tangentially through an entry port approximately halfway up the chamber wall so that it creates a swirling vortex flow pattern. The first flow field allows the solids to settle out by gravity; the second vortex that is generated in the skirt causes settleable solids to be concentrated at the bottom. The different configurations that have been developed are differentiated by the nature and type of internal flow modifying components and the location of inlets and outlets. The advantages of the vortex (or swirl) concentrator are that it includes no moving parts or external power requirement and has a high hydraulic loading, which results in compact size and low operational cost (Field and O'Connor, 1996). The successful application of swirl and vortex technologies in a stormwater treatment system depends on the following factors:

- consistent and appropriate flow measurement, wastewater sampling and characterization protocols,
- appropriate data management technique, especially the calculation efficiency,
- an understanding of swirl/vortex mechanisms, with realistic performance expectations, and
- appropriate application or placement in the treatment system (Field *et al.*, 1997).

Continuous Deflective Separator (CDS) appears to have similar features to the vortex separator, but the CDS design introduces a filtration mechanism for solids separation. The filtration mechanism, when combined with circular flow action and particle sedimentation, increases removal rates beyond vortex separators during times of high flows (Schwarz and Wells, 1999). The CDS unit works by deflecting the inflow and associated pollutants away from the main flow stream into a separation chamber. The chamber has a sump at the bottom and a screen in the upper section. The screen acts to remove the gross pollutants allowing the filtered water to pass through to a return system, while the floatable solids are kept in continuous motion on the water surface by incoming flow in the chamber, thus keeping the solids in the chamber from blocking the screen. With the heavy solids settling to the bottom of the sump, the CDS unit acts as a continuous cleaning unit, since the solids do not become imbedded in the filter screen as in a direct screening situation (Schwarz and Wells, 1999). The CDS mechanism is one of the new methods for removing suspended solids from surface runoff, and studies show that screen sizing has a significant influence on the sediment trapping efficiency of the CDS unit (Wong *et al.*, 1996).

Advantages:

- May provide the desired performance in less space and less cost compared to other BMPs.
- May be more cost-effective pre-treatment devices than traditional wet or dry basins.
- Mosquito control may be less of an issue than with traditional wet basins.

Limitations:

- It is likely that vortex separators are not as effective as wet vaults at removing fine sediments, on the order <50 to 100 μm in diameter.
- Area served is limited by the capacity of the largest models.
- The non-steady flows of stormwater decrease the efficiency of vortex separators from what may be estimated or determined from testing under constant flow.
- Do not remove dissolved pollutants; a loss of dissolved pollutants may occur as accumulated organic matter decomposes in the units.

2.3 Removal Processes Occurring in Structural BMPs

The high population of pollutant sources in urbanized areas contributes large quantities of pollutants that accumulate on streets, rooftops and other surfaces. During rainfall or snowmelt, these pollutants are mobilized and transported into the storm drain system, where they are conveyed and ultimately discharged to waterways. BMPs are implemented to remove these pollutants from runoff in order to reduce the impacts to receiving waters. A wide range of pollutants in urban runoff can be removed by properly designed, constructed and maintained structural BMPs through a number of physical and biochemical processes. The U.S. EPA identifies 115 organic compounds and 14 metals as "priority pollutants" in stormwater streams that can be removed by six main processes (Scholze *et al.*, 1993):

- adsorption on suspended solids followed by particle sedimentation or filtration,
- adsorption on an absorbing medium such as peat or other organic material,
- volatilization (common during aeration processes),
- photodegradation via photolysis,
- biodegradation (aerobic and anaerobic)/biological nitrification/de-nitrification, and
- phytoremediation/bioassimilation/biological uptake.

One or more of these treatment processes may occur in the treatment BMP systems to remove the pollutants of concern in stormwater runoff (Table 2-3).

Table 2-3. Pollutant Removal Mechanisms in Common Urban Stormwater BMPs

Mechanism	Pollutants Affected	Promoted by
Sedimentation	Solids, BOD, pathogens, COD (chemical oxygen demand), P, N, metals	Low turbulence
Filtration	Solids, BOD, pathogens, COD, P, N, metals	Fine, dense herbaceous plants
Sorption	Dissolved P, metals, synthetic organics	High soil Al, Fe; high soil organics, circum-neutral pH

Oxidation	COD, petroleum hydrocarbons, synthetic organics	Aerobic conditions
Volatilization	volatile petroleum hydrocarbons and synthetic organics	High temperature and air movement
Precipitation	Dissolved P, metals	High alkalinity
Biological Nitrification	NH_3 - N	Dissolved oxygen > 2.0 mg/L, low toxics, temp. > 5-7 C, circum-neutral pH
Microbial Decomposition	BOD, COD, petroleum hydrocarbons, synthetic organics	High plant surface area and soil organics
Phytoremediation	aromatics, chlorinated aliphatics, hydrocarbons, nutrients	rhizosphere microbial degradation, plant-produced enzymes

(Horner, 1995)

2.3.1 Sedimentation

Sedimentation is the removal of suspended particulates from the water column by gravitational settling. This can be a major mechanism of pollutant removal, particularly soil particles and TSS in BMPs such as ponds and constructed wetlands. The settling of discrete particles is dependent upon particle size and settling velocity; fluid density; fluid viscosity; particle diameter and shape; turbulence or short-circuiting; peak flowthrough rate; and volume of water (Stahre and Urbonas, 1990). Pollutants such as metals, hydrocarbons, nutrients and oxygen demanding substances can become adsorbed or attached to particulate matter, particularly clay soils; removal of stormwater particulates by sedimentation helps remove a large portion of pollutants associated with particulates from the water column. Stormwater BMPs that utilize settling are usually suited for dual purposes; that is, they can also provide storage volume for peak rate control, channel erosion and/or flood control (VA DCR, 1999).

The main factor governing the efficiency of a BMP at removing suspended matter by sedimentation is the time available for particles to undergo settling. Numerous research investigations show that significant settling of urban pollutants occurs in the first 6 to 12 h of detention, and fine particulates such as clay and silt require detention times of days or even weeks to settle out of suspension (Clar *et al.*, 2003). Evaluating the settling characteristics of the particulates in runoff becomes a prerequisite to designing a BMP in order to determine the detention time necessary for adequate settling to occur. Settling of particulates is also dependent upon the initial concentration of suspended solids (SS) in the runoff; runoff with higher initial concentrations of SS will have a greater removal efficiency.

The ability of pollutants to settle is directly influenced by particle size. Smaller particles take longer to settle, and conversely, the larger the particle, the faster is its settling velocity. However, particle size is not the only factor governing settling ability. The settling ability also depends on the difference between the density of the fluid suspending the particle and the density of the particle. Large, dense particles such as sand will fall through fluid at a faster rate than smaller, less dense particles, such as clay. The volume of particles suspended within the fluid also governs the settling rate. The more particles are suspended within the fluid, the faster the rate of settling, but at some point the rate of settling will bottom out (Clar *et al.*, 2003).

Turbulence, eddies, multilayered flows, circulation currents, and diffusion at inlets and outlets affect the settling ability of particles as each of these factors can resuspend particles into the water column. A decrease in flowthrough rate and surface loading improves sedimentation with the most significant difference observed for larger particles. Actual field conditions must take into account the particle settling velocity and surface loading rates during runoff conditions as sediment removal under these conditions varies with storm intensity. The size of the body of water relative to stormwater runoff will also determine the settling ability of sediment. The larger the stormwater loading rate, the lower the removal of sediment by settling. Settling also occurs after stormwater is trapped and ponded between storms. Because the intervals between storm events occur randomly, understanding the effective ratio of storage volume to mean runoff rate and the ratio of sediment volume removed to mean runoff rate is essential to predicting long-term averages (Clar *et al.*, 2003).

The most widely used stormwater management practices that employ sedimentation are retention and detention structures such as ponds and constructed wetlands. These can be designed to effectively remove sediment from stormwater. Stormwater management basins with a permanent pool of water have a removal percentage of total suspended solids of about 50 to 90% (MWCOG, 1983; OWML, 1983; Dorman *et al.*, 1989; City of Austin, 1990). Extended detention ponds have a similar percentage of removal (MWCOG, 1983; City of Austin, 1988; City of Austin, 1990). However, detention ponds may have a lower sediment removal efficiency over the long-term than retention ponds, as seen from a few investigations (Clar *et al.*, 2003). This is because of the resuspension of sediments deposited on the detention pond bed from previous storm events, during a new storm event.

Sedimentation removes particulates and other dissolved materials adsorbed to settleable particles in a structural BMP, yet, the removal rate by settling of pollutants other than sediment particles is inconclusive. One of the reasons for this confusion is a lack of understanding regarding the specific process(es) in a BMP responsible for removing the pollutant(s) of concern. In retention ponds, for example, processes such as settling, biological uptake, volatilization, infiltration to groundwater, and adsorption occur simultaneously. While nitrogen, phosphorus, and bacteria may be removed to some extent by adsorption to larger particulates, this is not expected to be a primary mechanism for their treatment. Metals, however, are present in particulate and dissolved form and some metal species are removed by coagulation and sedimentation.

2.3.2 Filtration

Filtration is typically limited to BMPs that address water quality and involves the removal of particulates by passing stormwater through a porous medium to strain the pollutants out of stormwater. Commonly used media in stormwater BMPs include soil, sand, gravel, peat, compost and various combinations such as peat/sand, soil/sand and sand/gravel. Existing media filtration practices commonly use trenches filled with sand or peat. Since the stormwater must pass through the filter media in order to be treated, these structures are limited to small drainage area (less than 5 acres) and low flow rates. A drawback to these structures is the overflow or bypass of large flows from high intensity storms. Filtration is a complex process that depends on a number of variables, including particle shape and size, void size in filter media, and velocity of the fluid moving through the media. Solids and associated pollutants such as metals and nutrients could be removed by filtration. Organic filtration media such as peat or leaf compost are also effective in removing soluble nutrients from urban runoff (Metro Council, 2001).

Typically, stormwater filters remove particulates and adsorbed pollutants, such as sediment, organic carbon, phosphorus, and many trace metals. Particulate pollutants are either trapped by cation/anion exchange or are prevented from moving beyond the filter. However, in some cases, the filtration process can increase the pollutant level of stormwater. The majority of N species associated with particulate matter are organic nitrogen compounds. Filters that inadvertently become anaerobic may cause nitrification of these species, and can release ammonia and nitrate into stormwater (Clark *et al.*, 2001). Once the treatment volume is achieved during a given storm, the excess runoff bypasses the filter and remains untreated. Filtration of infiltrated flow to remove sediments is more complicated, consisting of interception, straining, flocculation and sedimentation as the water percolates through the granular subsurface (Cammermayer *et al.*, 2000). The process is augmented as clay particles are adsorbed to positively charged organic matter, which can enhance settling in surface flows and improve retention in the subsurface. However, in vegetated stormwater channels, the sorption mass rate is quite low relative to sedimentation rates.

2.3.3 Sorption

The clay and organic matter in soil hold a negative charge. The ability of soil organic matter to hold cations such as calcium and aluminum represent the soil's cation exchange capacity. This process is most readily used to remove pollutants from stormwater. Organic matter such as peat or leaf matter in the filter media uses its cation exchange capacity to bind pollutants to the filter. The treatment of all runoff through filter media and biofilters, such as the bioretention cell (Clar and Green, 1993) are other examples of cation exchange processes. The media traps particulates (through filtration), adsorbs organic chemicals, and removes up to 90% of solids, 85% of oil and grease, and 82 to 98% of heavy metals (through cation exchange from leaf decomposition). While adsorption is not a common mechanism used in stormwater BMPs, it can still occur in infiltration systems where the underlying soils contain appreciable amounts of clay; in organic filters; or in wetland systems. Dissolved metals that are contained in stormwater runoff can be bound to the clay particles as stormwater runoff percolates through clay soils in infiltration systems.

The extent to which a given metal is adsorbed is affected by a number of factors, including competitive effects of other ionic metals; the presence of iron and manganese oxides and organic carbon; and, especially, pH (Clar *et al.*, 2003). Treatment trains that include adsorptive media may provide effective treatment for dissolved metals. Such media include compost, granulated activated carbon, or diatomaceous earth, all of which work on a cation exchange principle.

2.3.4 Phytoremediation

Plants are able to degrade (break down) organic pollutants through their metabolic processes, and the use of aquatic plants to treat wastewater, and the use of wetlands to treat farming effluent and mining runoff is well known (Yu and Nawang, 1993). Phytoremediation is an umbrella term that covers many different plant-based approaches for cleaning up contaminated environments, and refers to the use of plants to degrade, sequester and stabilize organic and metal pollutants in stormwater (U.S. EPA, 1998b). In simple terms, this means rendering pollutants harmless by using green plants to remove them from the environment. More recently, the bacterial activity associated with the roots of grasses and other plants has been explored for its organic degradation potential (enhanced rhizosphere phytodegradation). The efficiency of phytoremediation may vary depending on the depth of soil and the type and species of pollutants in water that are most available for plant uptake (U.S. EPA, 2001b).

Phytoremediation is a recent technology with immense development potential. The positive effects of plants can be both direct and indirect, and include:

- increased microbial degradation in the rhizosphere, including co-metabolism,
- uptake and accumulation in roots and foliage,
- degradation in the plant,
- volatilization of the compounds, and
- plant-produced enzymes which degrade pollutants (Rasmussen and Olsen, 2004).

Phytoremediation can involve any of the following approaches. Phytoextraction uses the ability of plants to take up and remove contaminants from soil and water and accumulate them in plant tissues, which may then be harvested and removed from the site. The use of plants and (or) their associated microbes to volatilize contaminants (volatile organic compounds, i.e., solvents, and recently, inorganics such as Hg, Se, etc.) from soil or water is known as phytovolatilization. In this process, plants take up water containing organic contaminants and release the contaminants or the breakdown products into the air through their leaves. Se volatilization appears to be a significant pathway of Se removal from contaminated sites (Terry, 2001b). Although transferring contaminants to the atmosphere may not achieve the goal of complete remediation, phytovolatilization may still be desirable in that it reduces prolonged soil exposure and the associated risk of groundwater contamination; another advantage is that there is no hazardous waste generation that warrants proper disposal measures as may be the case in phytoextraction. Phytodetoxification involves the ability of plants to change the chemical species of the contaminant to a less toxic form. Research shows the ability of plants to take up toxic Cr(VI) species and convert it into the non toxic trivalent Cr(III). Phytostabilization uses plants to immobilize contaminants chemically and physically at the site, thereby preventing their movement to surrounding areas (Terry, 2001a).

Phytotransformation, also referred to as phytodegradation, is the breakdown of organic contaminants sequestered by plants via (i) metabolic processes within the plant, or (ii) the effect of compounds, such as the enzymes deoxygenase and halogenase, which are produced by the plant. The organic contaminants are degraded into simpler compounds that are integrated with plant tissue, which in turn, foster plant growth. Remediation of a site by phytotransformation is dependent on the direct uptake of contaminants from the media and accumulation in the vegetation. Certain enzymes produced by plants are able to breakdown and convert chlorinated solvents (e.g., trichloroethylene), ammunition wastes, and herbicides. This technology can also be used to remove contaminants from petrochemical sites and storage areas, fuel spills, landfill leachates, and agricultural chemicals. Successful implementation of this technology requires that the transformed compounds that accumulate within the plant be non toxic or significantly less toxic than the parent compounds. Phytotransformation may also be used in concert with other remediation technologies or as a polishing treatment. For example, a combination of phytoremediation using orchard grass and a soil/sand filter material can efficiently treat creosote-contaminated groundwater (Rasmussen and Olsen, 2004).

The direct uptake of chemicals into plant tissue via the root system is dependent on uptake efficiency, transpiration rate, and concentration of the chemical in soil water. Uptake efficiency depends on chemical speciation, physical/chemical properties, and plant characteristics, whereas transpiration rate depends on plant type, leaf area, nutrients, soil moisture, temperature, wind conditions, and relative humidity. Two processes of remediation can occur after the organic compound has been translocated by the plant: (i) storage of the chemical and its fragments into the plant via lignification, and (ii) complete conversion to carbon dioxide and water. These techniques have been successfully employed to treat and remove the following contaminants in the environment: aromatics (BTEX); chlorinated aliphatics (TCE); herbicides (atrazine, alachlor); hydrocarbons (TPH); nutrients (NO_3^-, NH_4^+, PO_4^{3-}) (U.S. EPA, 1998b).

2.3.5 Biological Processes

Biological processes are one of the most effective types of removal mechanisms for soluble pollutants such as nutrients in stormwater runoff and an important mechanism of nutrient control in stormwater BMPs (VA DCR, 1999). Vegetated lands help prevent erosion and remove contaminants from surface runoff in a process known as bioassimilation. It also helps to slow down the runoff, giving it more time to infiltrate into the soil. A combination of shallow permanent pool depths and abundant vegetation help to create conditions that allow a natural food chain to develop; marsh plants, algae and microorganisms that grow on the shallow organic rich sediments take up soluble forms of nutrients needed for their growth. Periodic harvesting of this vegetation allows for permanent removal of these nutrients. BMPs suited for this pollutant removal mechanism include enhanced ED as well as retention ponds, constructed stormwater wetlands, and in some cases bioretention. Some wetland plants not only assimilate contaminants; wetland plants such as cattails have been shown to assimilate, and in some cases, transform pathogenic bacteria and metals (1994). The use of submerged aquatic vegetation as the dominant vegetation in treatment wetlands in south Florida results in a higher P removal performance than wetlands dominated by rooted, emergent plants, with the newly accreted sediment being the likely, long-term sink for this P (Knight *et al.*,

2003). Macrophytes reduce flow velocity, increase sedimentation and retention by deposition in running waters and contribute to total monthly phosphorus retention (up to 25%) by increasing deposition of particulate organic matter (Schulz *et al.*, 2003).

Many stormwater BMPs utilize a combination of these pollutant removal mechanisms. In some cases, development of an organic layer occurs within a BMP that has been in operation for a period of time, thereby increasing the adsorption potential of the BMP. BMPs that include plants and grasses also display increased pollutant removal efficiency over time as the biomass increases. This increase in biomass slows the velocity of the runoff through the BMP and allows for increased gravitational settling and filtering of pollutants, as well as decreased export of sediment and attached pollutants via erosion (VA DCR, 1999).

2.4 Performance of Structural BMPs

The pollutant removal capability of a BMP is primarily governed by three inter-related factors: the removal mechanisms used; the fraction of the annual runoff volume that is effectively treated; and, the nature of the urban pollutant being removed. The nature of the pollutant being removed often sets an upper limit on the potential removal rate that can be achieved. While particulate pollutants such as sediment and lead are relatively easier to remove by common BMP removal mechanisms including settling and filtering, it is not the case with soluble pollutants such as nitrate, phosphates and some trace metals. The above-mentioned removal mechanisms have little or no effect and require the additional use of biological methods such as uptake by bacteria, algae, rooted aquatic plants or terrestrial vegetation. Most BMPs can achieve an extremely high removal rate for suspended sediment and trace metals that exist largely in particulate forms. However, removal rates that are much lower are generally obtained for total phosphorus, oxygen-demanding materials, and total nitrogen, since they typically exist in both particulate and soluble forms. Tables 2-4 summarizes the processes governing pollutant removal in commonly used structural BMPs.

Table 2-4. Processes Governing Pollutant Removal in Commonly Used Structural BMPs

Pollutant Constituents	**Structural BMP Type and Process Mechanisms**				
	Pond	**Wetland**	**Biofilters**	**Infiltration**	**Sand Filter**
Heavy metals	Sorption Settling	Sorption Settling Phyto-remediation	Sorption Filtration	Sorption Filtration Phyto-remediation Settling	Sorption Filtration

Toxic Organics	Sorption Bio-degradation Settling Phyto-volatilization	Sorption Bio-degradation Settling Phyto-volatilization	Sorption Filtration	Sorption Filtration Settling Phyto-volatilization	Sorption Filtration
Nutrients	Bio-assimilation	Bio-assimilation Phyto-remediation	Sorption	Sorption Bio-assimilation Phyto-remediation	Sorption
Solids	Settling	Sorption Settling	Sorption Filtration	Sorption Filtration Settling	Filtration
Oil & Grease	Sorption Settling	Sorption Settling	Sorption	Sorption Settling	Sorption
BOD5	Bio-degradation	Bio-degradation	Bio-degradatio n	Bio-degradation	Bio-degradation
Pathogens	Settling UV irradiation	UV irradiation* (sunlight) Sedimentation Aggregation Oxidation Antibiosis	Filtration	Filtration Settling	Filtration Predation

(Scholze *et al.*, 1993) *(Davies and Bavor, 2000)

The efficiency of a given BMP in removing pollutants is dependent upon a number of site-specific variables: (i) size, type and design of the BMP; (ii) soil type and characteristics; (iii) geology and topography of the site; (iv) intensity and duration of rainfall; (v) length of antecedent dry periods; (vi) climatological factors such as temperature, solar radiation, and wind; (vii) size and characteristics of the contributing watershed; and, (viii) properties and characteristics of the various pollutants (U.S. EPA, 1999b). Urban stormwater BMP performance is affected by various factors as shown in Table 2-5.

Table 2-5. Factors affecting the Performance of Structural BMPs

Design Characteristics	Detention time Storage volume Shape of BMP (length : width ratio of detention pond, slope, flow path characteristics, buffer width) Pond surface area relative to contributing catchment area Type of substrate/sediment Auxiliary devices (e.g., baffles, sediment forebay) Presence of vegetation
Processes	Sedimentation Filtration Adsorption Volatilization Oxidation Precipitation Biological nitrification Microbial decomposition Phytoremediation
Environmental Conditions	Storm intensity, loading rate Drainage basin land cover Physical-chemical properties of BMP surface water Particle size distribution and settling velocity Pre-storm water quality in BMP Thermal stratification

(Law and Band, 1998)

2.4.1 Suspended Solids, Nutrients, and Heavy Metals, and BMP Performance

BMPs are primarily designed to remove TSS and pollutants sorbed to particles, with gravitational settling being the predominant process for pollutant removal (Schueler *et al.*, 1992). The ability to treat TSS is a function of particle size distribution, storm intensity, loading rate, and geometry and age of BMP facility. An increase in TSS concentration observed with increased storm intensity could be due to larger size fractions of TSS, that result in a greater removal efficiency (Ferrara and Witkowski, 1983). Pond geometry also influences the sedimentation rate in that the finer particles do not settle out when the length-to-width ratio is insufficient; in detention ponds draining a commercial complex, observations of sediment sorting showed coarsest particles settling nearest the inlet (Marsalek *et al.*, 1997). The effectiveness of sedimentation is further limited by the available storage volume and hydraulic loading rate of runoff to the BMP. BMP performance is adversely affected under the following conditions:

- if the runoff volume is greater than the storage volume of the BMP,
- short-circuiting within the pond if sediment is allowed to accumulate and consequently reduce storage volume (U.S. EPA, 1993b), and
- age of the BMP.

In order to improve stormwater quality when the water is polluted with excessive nutrient levels, BMP facilities are designed to provide sedimentation for particle-phase nutrients and biological uptake for soluble nutrients (Martin, 1988). Wet detention ponds and constructed wetlands act as a sink, source or transformer of nutrients. However, the performance of BMPs for nutrient-enriched stormwater is unpredictable. This could be due to the complex nature of nutrient speciation, which is affected by seasonality, detention time, organic matter content, oxygen availability and plant biomass. The most effective phosphorus removal mechanisms are adsorption, complexation, precipitation reactions with Al, Fe, Ca and clay particles, and by peat accretion. Pond conditions amenable to nitrification-denitrification processes are most effective for nitrogen removal. Similar to TSS, removal efficiencies of nutrients in BMPs are affected by particle size, quality of the substrate and age of the treatment facility. BMP performance is highly variable for nutrients compared to other constituents with regard to both particulate and dissolved phases, as well as different species of nutrients (Tanner *et al.*, 1997). Also, the soluble phase of nutrients may not be effectively treated in a wet detention basin. Studies on wet detention facilities show no clear relationship between particle size distribution and storm intensity to explain the variability of reduction for all particle size classes for total Kjeldahl nitrogen (TKN) and total phosphorus (TP). The net export of pollutants may be attributed to the large proportion of pollutants associated with the soluble size fraction of solids (i.e., $< 1\mu m$) (Law and Band, 1998).

BMPs generally contribute to the reduction of total heavy metal pollutants from urban stormwater. Overall, studies report a moderate to high removal efficiency for wet detention basins, constructed wetlands, and combined pond-wetland systems (Schueler *et al.*, 1992). It is assumed that pollutants in urban stormwater are largely sorbed to suspended sediments and therefore sedimentation is considered to be the predominant removal mechanism for heavy metals. The concentration of heavy metals in sediments indicates the improvement that BMPs have on stormwater quality (Marsalek *et al.*, 1997). Despite the strong ability of particulates and organic matter to sequester heavy metals, TSS are not good surrogates for other pollutants (Law and Band, 1998). This however, counters a design principle for BMPs which assumes that an 80% removal of TSS will control other pollutants, including heavy metals (U.S. EPA, 1993b). Illustrating the presence of heavy metals in urban runoff in the dissolved phase implies a low reliance on sedimentation to remove heavy metals from the water column. Adverse effects induced by sediment contamination may counter the benefits derived from improved water quality. The solubility of heavy metals is affected by cation-exchange capacity, sensitivity to pH and Eh, which in turn affects the ability of sediments to sequester heavy metals from the water column. Heavy metal mobilization from sediments can be enhanced due to intermittent flooding, which produces alternating periods of aerobic and anaerobic conditions. Also, the change in the oxidized state of an environment can alter the speciation of metals, and hence their bioavailability.

REFERENCES

Andoh, R.Y.G. and A.J. Saul. (2002). The Use of Hydrodynamic Vortex Separators and Screening Systems to Improve Water Quality. In: *3rd International Conference on Sewer Processes and Networks Proceedings*, Paris, France.

Atlanta Regional Commission. (2001). *Georgia Stormwater Management Manual Volume 2: Technical Handbook.* Atlanta Regional Commission, www.georgiastormwater.com

Bradner, J.N. and M.P. Wanielista. (1992). *Project SMART.* Florida Department of Environmental Regulation, Orlando, FL.

State of California Department of Transportation (Caltrans). (2002). *Proposed Final Report: BMP Retrofit Pilot Program.* CTSW-RT-01-050. California Department of Transportation Report CTSW-RT-01-050 , and Sacramento, CA.

Cammermayer, J.W., R.R. Horner, and N. Chechowitz. (2000). *Vegetated Stormwater Facility Maintenance. Research Report.* Prepared for Washington State Transportation Commission, Department of Transportation (WSDOT) in cooperation with U. S. Department of Transportation, Federal Highway Administration (FHWA), Seattle, WA.

California Stormwater Quality Association (CASQA). (2003). *Stormwater Best Management Practice Handbook: New Development and Redevelopment.* www.cabmphandbooks.com.

City of Austin. (1988). *Environmental Criteria Manual.* Environmental Resource Management Division, Austin, TX.

City of Austin. (1990). *Removal Efficiencies of Stormwater Control Structures. Final Report.* Austin, TX.

Clar, M., B.J. Barfield, and S. Yu. (2003). *Considerations in the design of treatment best management practices (BMPs) to improve water quality.* EPA 600/R-03/103. U.S. Environmental Protection Agency, Office of Research and Development, Cincinnati, OH. www.epa.gov/ORD/NRMRL/Pubs/600r03103/600r03103.htm October 2003.

Clar, M.L. and R. Green. (1993). *Design Manual for Use of Bioretention in Stormwater Management.* Department of Environmental Resources Watershed Protection Branch, Prince George's County, MD.

Clark, S., R. Pitt, and P. Brown. (2001). Effect of Anaerobiosis on Filter Media Pollutant Retention. In: *Information and Monitoring Needs for Evaluating the Mitigating Effects of BMPs - Engineering Foundation and the American Society of Civil Engineers Conference*, Snowmass, CO.

Center for Watershed Protection (CWP). (1998). *Better Site Design: A Handbook for Changing Development Rules in Your Community.* Ellicott City, MD.

CWP. (2000). An Introduction to Better Site Design. In: *The Practice of Watershed Protection*, Schueler, T.R. and H.K. Holland, (editors). Center for Watershed Protection,, Ellicott City, MD. Pages 623-632.

Davies, C.M. and H.J. Bavor. (2000). The fate of stormwater-associatied bacteria in constructed wetland and water pollution control pond systems. *Journal of Applied Microbiology* **89**: 349-360.

D'Itri, F.M., (editor). (1992). *Chemical Deicers and the Environment.* Lewis Publishers, Inc., Chelsea, MI.

Dorman, M.E., J. Hartigan, R.F. Steg, and T. Quasebarth. (1989). *Retention, detention and overland flow for pollutant removal from highway stormwater runoff. Volume I. Report No. FHWA/RD 89/202*. Federal Highway Administration, Washington, D. C.

Ferrara, R. and P. Witkowski. (1983). Stormwater quality characteristics in detention basins. *Journal of Environmental Engineering* **190**(2): 428-447.

FHWA. (2000). *Stormwater Best Management Practices in an Ultra-Urban Setting: Selection and Monitoring*. U.S. Department of Transportation, Washington, D. C. http://www.fhwa.dot.gov/environment/ultraurb/index.htm

Field, R. and T.P. O'Connor. (1996). Swirl Technology: Enhancement of Design, Evaluation and Application. *Journal of Environmental Engineering* **122**(8): 741-748.

Field, R., D. Averill, T.P. O'Connor, and P. Steel. (1997). Vortex Separation Technology. *Water Quality Research Journal of Canada* **32**(1): 185-214.

Field, R. and D. Sullivan, (editors). (2003). *Wet-Weather Flow in the Urban Watershed: Technology and Management*. CRC Press LLC, Boca Raton, FL.

Florida DER. (1988). *The Florida Development Manual: A Guide to Sound Land and Water Management*. Florida Department of Environmental Regulation, Tallahassee, FL.

GreenRoofs. Exploring the Ecology of Organic Green Roof Architecture, www.greenroofs.com.

Greater Vancouver Sewerage and Drainage District (GVSDD). (1999). *Best Management Practices Guide for Stormwater*. Burnaby, B. C.

Horner, R. (1995). *Constructed Wetlands for Urban Runoff Water Quality Control*. Department of Civil Engineering, University of Washington, Seattle, WA.

Knight, L.R., B. Gu, A.R. Clark, and M.J. Newman. (2003). Long-term phosphorus removal in Florida aquatic systems dominated by submerged aquatic vegetation. *Ecological Engineering* **20**: 45-63.

Konieek, Z., K. Pryl, and M. Suchanek. (1996). Practical Applications of Vortex Flow Separators in the Czech Republic. *Water Science and Technology* **33**(9): 253-260.

Law, N.L. and L.E. Band. (1998). Performance of urban stormwater best management practices, http://www.unc.edu/depts/geog/them/projects/BMP.html. University of Carolina - Chapel Hill, Department of Geology, Chapel Hill, NC.

Lee, J., K. Bang, J. Choi, L.H. Ketchum Jr, and Y. Cho. (2003). The vortex concentrator for suspended solids treatment. *Water Science and Technology* **47**(9): 335-341.

Low Impact Development Center (LIDC). *Green Roofs*. Beltsville, MD. http://www.lid-stormwater.net/greenroofs/greenroofs_home.htm.

Liptan, T. and E. Strecker. (2003). EcoRoofs (Green Roofs) - A More Sustainable Infrastructure. In: *National Conference on Urban Stormwater: Enhancing Programs at the Local Level. Proceedings. EPA-625-R-03-003*, February 17-20, Chicago, IL. http://www.epa.gov/owow/nps/natlstormwater03/20Liptan.pdf.

Marsalek, J., W.E. Watt, B.C. Anderson, and C. Jaskot. (1997). Physical and chemical characteristics of sediment from a stormwater management pond. *Water Quality Research Journal of Canada* **32**(1): 89-100.

Martin, E.H. (1988). Effectiveness of an urban runoff detention pond - wetland system. *Journal of Environmental Engineering* **114**(4): 810-827.

Maryland Department of the Environment (MDE). (2000). *Maryland Stormwater Design Manual Volumes I & II*. Center for Watershed Protection and Maryland Department of the Environment, Baltimore, MD. http://www.mde.state.md.us.

Michigan Department of Environmental Quality (MDEQ). (1998). *Guidebook of Best Management Practices for Michigan Watersheds*. Surface Water Quality Division.

(Metro Council). (2001). *Minnesota Urban Small Sites BMP Manual. Stormwater Best Management Practices for Cold Climates*. Prepared for the Metropolitan Council by Barr Engineering Company,, St. Paul, MN.

Metropolitan Washington Council of Governments (MWCOG). (1983). *Urban Runoff in the Washington Metropolitan Area - Final Report Washington D. C. Area Urban Runoff Project*. Prepared for U.S. EPA and WRPB, Washington, DC.

MWLAP. (1992). Best Management Practices to Protect Water Quality, http://wlapwww.gov.bc.ca/wat/wq/nps/BMP_Compendium/BMP_Introduction/bmphome .htm. Ministry of Water, Land and Air Protection, Government of British Columbia,.

Northern Virginia Planning District Commission (NVPDC). (1979). *Guidebook for Screening Urban Nonpoint Pollution Management Strategies*. Annandale, VA.

Northern Virginia Planning District Commission (NVPDC). (1996). *NonStructural Urban BMP Handbook: A guide to nonpoint source pollution prevention and control through nonstructural measures*. Annandale, VA.

Northern Virginia Planning District Commission and Engineers and Surveyors Institute (NVPDC and ESI). (1992). *Northern Virginia BMP Handbook: A guide to planning and designing best management practices in northern Virginia*. Annandale, VA.

New York State Department of Environmental Conservation (NYSDEC). (2001). *New York state stormwater management design manual (Draft)*. Division of Water, Albany, NY. http://www.dec.state.ny.us/website/dow/swmanual. October 2001.

Occoquan Watershed Monitoring Laboratory (OWML). (1983). *Final Contract Report: Washington Area NURP Project*. Prepared for Metropolitan Washington Council of Governments, Manassas, VA.

Prince George's County Department of Environmental Resources (PGDER). (1993). *Design Manual for Use of Bioretention in Stormwater Management*. Division of Environmental Management, Watershed Protection Branch, Landover, MD.

Pierce County Public Works and Utilities (Pierce County). *Stormwater Pollution Prevention Manual - A Guide to Best Management Practices for Industries, Businesses and Homeowners*. University Place, WA. http://www.co.pierce.wa.us/pc/services/home/environ/water/swm/sppman/bmps9.htm.

Pitt, R., B. Robertson, P. Barron, A. Ayyoubi, and S. Clark. (1999). *Stormwater Treatment at Critical Areas: The Multi-Chambered Treatment Train (MCTT). EPA-600-R-99-017*. National Risk Management Research Laboratory, United States Environmental Protection Agency, Cincinnati, OH.

Prince George's County (DER). (2002). *Low Impact Development Integrated Management Practices Guide Book*. government in Prince George's County, Maryland,

Rasmussen, G. and R.A. Olsen. (2004). Sorption and biological removal of creosote-contaminants from groundwater in soil/sand vegetated with orchard grass (Dactylis glomerata). *Advances in Environmental Research* **8**: 313-327.

Scholze, R., V. Novotny, and R. Schonter. (1993). Efficiency of best management practices for controlling priority pollutants in runoff. *Water Science and Technology* **28**(3-5): 215-224.

Schueler, T.R. (1987). *Controlling Urban Runoff: A practical manual for planning and designing urban BMPs*. Metropolitan Washington Council of Governments, Washington, DC.

Schueler, T.R. (1992). *Design of Stormwater Pond Systems*. Metropolitan Washington Council of Governments,, Washington, D. C.

Schueler, T.R., P.A. Kumble, and M.A. Hearty. (1992). *A Current Assessment of Urban Best Management Practices: Techniques for reducing nonpoint source pollution in the coastal zone*. Metropolitan Washington Council of Governments, Washington, DC.

Schulz, M., H.-P. Kozerski, T. Pluntke, and K. Rinke. (2003). The influence of macrophytes on sedimentation and nutrient retention in the lower river Spree (Germany). *Water Research* **37**: 569-578.

Schwarz, T. and S. Wells. (1999). Stormwater Particle Removal using Cross-Flow Filtration and Sedimentation. In: *Advances in Filtration and Separation Technology Volumes 13 A & B*, Leung, W., Ptak, T.,, (editor). American Filtrations and Separations Society. Pages 219-226.

Stahre, P. and B.R. Urbonas. (1990). *Stormwater Detention for Drainage, Water Quality and CSO Control*. Prentice Hall, Englewood Cliffs, NJ.

Strecker, E. (2001). Low-Impact Development (LID) Is it really low or just lower? In: *Linking Stormwater BMP Designs and Performance to Receiving Water Impact Mitigation - Proceedings of an Engineering Foundation Conference*, Urbonas, B.R., (editor). American Society of Civil Engineers, Snowmass Village, CO. Pages 210-222.

Tanner, C.C., J.P.S. Sukias, and M. Upsdell. (1997). Relationships between loading rates and pollutant removal during maturation of gravel-bed constructed wetlands. *Journal of Environmental Quality* **27**: 448-458.

Taylor, A. and T. Wong. (2002). *Nonstructural Stormwater Quality Best Management Practices - An Overview of Their Use, Cost and Evaluation. Technical Report 02/11.* Cooperative Research Center for Catchment Hydrology, Victoria, NZ.

Terry, N. (2001a). http://www.hawaii.edu/abrp/Technologies/phytran.html).

Terry, N. (2001b). *Use of flow-through constructed wetlands for the removal of selenium in agricultural-tile drainage water. Annual Report 2000-2001.* U.C Salinity/Drainage Task Force, Division of Agriculture and Natural Resources, University of California, Berkeley, CA.

United States Environmental Protection Agency (U.S. EPA). (1977). *Swirl Device for Regulating and Treating Combined Sewer Overflows. EPA Technology Transfer Capsule Report. EPA-625-2-77-012.* Washington, DC.

United States Environmental Protection Agency (U.S. EPA). (1983). *Results of the Nationwide Urban Runoff Program. Volume I - Final Report. EPA-832-R-83-112.* Washington, DC.

United States Environmental Protection Agency (U.S. EPA). (1992). *Stormwater Management for Construction Activities: Developing Pollution Prevention Plans and Best Management Practices. EPA-832-R-92-005.* Washington, DC.

United States Environmental Protection Agency (U.S. EPA). (1993a). *Guidance Specifying Management Measures for Sources of Nonpoint Pollution in Coastal Waters. EPA 840-B-92-002.* . Washington, DC.

United States Environmental Protection Agency (U.S. EPA). (1993b). *Handbook - Urban Runoff Pollution Prevention and Control Planning. EPA-625-R-93-004.* Washington, DC.

United States Environmental Protection Agency (U.S. EPA). (1993c). *Investigation of Inappropriate Pollutant Entries into Storm Drainage Systems EPA-600-R-92-238*. Washington, DC.

United States Environmental Protection Agency Region 5 (U.S. EPA). (1998a). *Illegal Dumping Prevention Guidebook*. Chicago, IL.

United States Environmental Protection Agency (U.S. EPA). (1998b). *A Citizen's Guide to Phytoremediation. EPA-542-F-98-011*. Office of Solid Waste and Emergency Response, Washington, DC.

United States Environmental Protection Agency (U.S. EPA). (1999a). *Stormwater Management Fact Sheet. EPA-832-F-99-001 through EPA-843-F-99-050.* Washington, DC.

United States Environmental Protection Agency (U.S. EPA). (1999b). *Preliminary Data Summary of Urban Stormwater Best Management Practices. EPA-821-R-99-012*. Office of Water, Washington, DC.

United States Environmental Protection Agency (U.S. EPA). (2000a). *Low Impact Development Design Strategies: An integrated design approach.* Prepared by Tetra Tech Inc., Fairfax, VA for Department of Environmental Resources, Prince George's County MD, funding provided by the United States Environmental Protection Agency, Washington, DC.

United States Environmental Protection Agency (U.S. EPA). (2000b). *Low Impact Development Hydrologic Analysis.* Prepared by Tetra Tech Inc., Fairfax, VA for Department of Environmental Resources, Prince George's County MD, funding provided by the United States Environmental Protection Agency, Washington, DC.

United States Environmental Protection Agency (U.S. EPA). (2000c). *Vegetated Roof Cover, Philadelphia, Pennsylvania. EPA-841-B-00-005D*. Washington, DC.

United States Environmental Protection Agency (U.S. EPA). (2000d). *Low Impact Development (LID) - A Literature Review. EPA-841-B-00-005*. Washington, DC.

United States Environmental Protection Agency (U.S. EPA). (2001a). *National Menu of Best Management Practices for Stormwater Phase II*. Washington, DC. http://cfpub.epa.gov/npdes/stormwater/menuofbmps/menu.cfm.

United States Environmental Protection Agency (U.S. EPA). (2001b). *A Citizen's Guide to Phytoremediation. EPA-542-F-01-002*. Washington, DC. http://www.epa.gov/superfund/sites.

United States Environmental Protection Agency (U.S. EPA). (2003). *National Conference on Urban Stormwater: Enhancing Program at the Local Level Proceedings Chicago, IL February 17-20, 2003. EPA-625-R-03-003.* National Risk Management Laboratory, Office of Research and Development, Cincinnati, OH.

UNEP IETC. (2000). Environmentally Sound Technologies in waste water treatment for the implementation of the UNEP Global Programme of Action (GPA) "Guidance on Municipal Wastewater" in collaboration with Murdoch University Environmental Technology Centre, http://www.unep.or.jp/ietc/publications/freshwater/sb_summary/index.asp. United Nations Environment Programme International Environmental Technology Centre.

University of Rhode Island. (1994). Technical Analysis Report. Prepared for The Saugatucket River Heritage Corridor Coalition by the Advanced Planning Studio of the Graduate Curriculum in Community Planning and Area Development, http://saugatucket.org/docs/SaugTechRpt/Saugatucket_Tech_Report.pdf.

Commonwealth of Virginia Department of Conservation and Recreation (VA DCR). (1999). *Virginia Stormwater Management Handbook, Volume I,*. Division of Soil and Water Conservation., Richmond, VA.

Washington State Department of Ecology (WA DOE). (1991). *Stormwater Management Manual for the Puget Sound Basin - Public Review Draft*. Olympia, WA.

Washington State Department of Ecology (WA DOE). (2001). *Stormwater Management Manual for Western Washington. Volume V - Runoff Treatment BMPs*. Water Quality Program, http://www.ecy.wa.gov/programs/wq/stormwater/index.html.

Walker, T.A. and T.H.F. Wong. (1999). *Effectiveness of Street Sweeping for Stormwater Pollution Control, Technical Report 99/08*. Cooperative Research Center for Catchment Hydrology, Melbourne.

Wanielista, M.P. and J.N. Bradner. (1992). *Project SMART - A Restoration Demonstration*. University of Central Florida, Orlando, FL.

Water Environment Federation and American Society of Civil Engineers (WEF and ASCE). (1998). *Urban runoff quality management*. 1-57278-039-8 and 0-7844-0174-8. Water Environment Federation, Alexandria, VA.

Wong, T.H.F., R.M. Wootton, and D. Fabian. (1996). A Solid Separator Using a Continuous Deflective System. In: *Proceedings 7th International Conference on Urban Stormwater Drainage*, Hannover, Germany. IAHR/IAWQ Joint Committee on Urban Storm Drainage.

Yu, S.L. and W.M. Nawang. (1993). Best Management Practices for Urban Stormwater Runoff Control. In: *Integrated Stormwater Management*, Field, R., M.L. O'Shea, and K.K. Chin, (editors). Lewis Publishers, Boca Raton, FL. Pages 191-206.H

3 Structural BMP Design Practices

Swarna Muthukrishnan, Richard Field and Daniel Sullivan

CONTENTS

3.1 INTRODUCTION

The selection and successful design of structural BMPs for stormwater quality enhancement is the cornerstone of stormwater management in newly developing and redeveloping urban areas. The cost effectiveness of each control has to be considered and measured against the actual environmental benefits realized. Design objectives for BMPs can be stated in terms of technology (e.g., by specifying a particular control device) or in terms of quantitative effect (e.g., by specifying a required degree of control or a maximum allowable effect). Quantitative objectives can be defined for both hydrological parameters and constituent removal performance parameters. Some examples of objectives based on hydrological parameters include peak flow rate and retention of a defined water volume for a specific period of time, while objectives based on chemical parameters include percent removal of specific chemical constituents and effluent concentration or mass discharge targets (WEF and ASCE, 1998).

In 1990, the U.S. EPA promulgated phase I of the NPDES (National Pollutant Discharge Elimination System) program regulations for stormwater discharges from municipal separate storm sewer systems (MS4s) (U.S. EPA, 1996), which required municipalities to reduce pollutants in urban runoff to the maximum extent practicable (MEP). The definition of MEP for the control of stormwater pollutant discharges has focused primarily on the application of economically achievable management practices. Stormwater runoff rates and volumes vary highly between storms; hence, the statistical probabilities of runoff events and their management have to be considered in developing practices to meet the MEP goal. It is therefore imperative to examine the hydrology of urban runoff and the type and size of storm runoff events to result in a robust design that has a high probability of meeting these regulations (WEF and ASCE, 1998).

Stormwater management programs were developed by various states in the 1980s when the prevailing outlook was that the quality of receiving streams could be sustained by controlling the flooding caused by increases in runoff volume from new development. The objective of these stormwater management facilities therefore was to control peak flows. Efforts on stormwater management during this time addressed BMP design for flood control based on hydrological procedures with little or no satisfactory guidelines or criteria set forth for water quality management. Most of these designs were flow-based, with an emphasis on the reduction of post-construction flows of the 2- and 10-yr storm events to pre-development levels in new development; this peak management approach addressed only stream channel erosion concerns besides providing adequate flood control in receiving waters and not water quality. However, changes in land use patterns due to increasing urbanization has resulted in a large increase in impervious cover and runoff, strongly highlighting the need to control both the quality and quantity of runoff in order to prevent stream channel erosion. In response to the provisions of the Clean Water Act, (originally enacted in 1977 with subsequent revisions), a number of activities such as the Nationwide Urban Runoff Program (NURP) (U.S. EPA, 1983) were initiated to characterize and quantify the water quality impacts of WWF, and municipalities began adopting BMPs for pollutant removal.

A growing national awareness of the wide range of environmental impacts of runoff and urbanization has resulted in BMPs being designed to control larger storms as well as smaller, but more frequent storms to achieve additional ecological benefits that include stream channel protection and restoration, groundwater infiltration, protection of riparian habitat and biota, and minimized thermal impacts. Collected runoff has also been used for irrigation, toilet flushing, and other non-potable purposes, including ponds and wetlands that also enhance urban aesthetics. This redirected approach in considering a watershed in its entirety and using BMPs for water quality improvement has led to several procedures being established to achieve removal of pollutants from storm runoff. A few approaches include the mandatory requirement to remove 80% of total suspended solids (TSS) from new development in coastal zone states (Coastal Zone Management Act (CZMA), 1972), and controlling the first flush of pollutants associated with a storm, mandating the capture of the first 0.5 to 1 in. of runoff (typically generated in the first hours of the 1-yr storm).

3.2 BMP Design Criteria

3.2.1 Unified Stormwater Sizing Criteria

The objective of any stormwater design criteria is to protect receiving waters from adverse impacts associated with urban runoff. This goal can be successfully accomplished by adopting a unified approach to sizing stormwater BMPs, which is influenced by several factors. A few examples include local hydrological conditions, rainfall-runoff pattern, the type of BMP to be installed, the volume of stormwater that would be treated, degree of imperviousness, prevailing stormwater regulations to be adhered to etc. As an example, the guidelines proposed by the Maryland Department of Environment, which consists of five main quality characteristics of stormwater, is presented in Table 3-1 (MDE, 2000) and briefly explained below.

Table 3-1. Summary Example of Unified Stormwater Sizing Criteria

Sizing Criteria	Description
Water Quality Volume (WQ_v) (acre-ft)	$WQ_v = [(P)(R_v)(A)]/12$ P = 1.0 in. in Eastern Zone and 0.9 in. in Western Zone R_v = volumetric runoff coefficient A = Area in acres

Recharge Volume (Re_v) (acre-ft)	$Re_v = [(S)(R_v)(A)]/12$ S = Soil Specific Recharge Factor Re_v is a sub-volume of WQ_v
Channel Protection Storage Volume (Cp_v)	Cp_v = 24 h (12 h in USE III and IV watersheds) extended-detention of the post-developed 1-yr 24 h storm event. Not required for direct discharges to tidal waters and the Eastern Shore of Maryland.
Overbank Flood Protection Volume (Q_{px})	Local review authorities may require that the peak discharge from the 10-yr storm event be controlled to the pre-development rate (Q_{p10}). No control of the 2-yr storm event (Q_{p2}) is required.
Extreme Flood Volume (Q_f)	Consult with the appropriate local reviewing authority. Normally no control is needed if development is excluded from the 100-yr flood plain and downstream conveyance is adequate.

(MDE, 2000)

Water Quality Volume (WQ_v)

Water quality volume is the storage needed to capture and treat the runoff from 90% of average annual rainfall. Numerically this is equivalent to an inch of rainfall multiplied by the volumetric runoff coefficient (R_v) and site area. Treatment of the WQ_v shall be provided at all developments where stormwater management is required. According to (MDE, 2000), a minimum WQ_v of 0.2 in./acre shall be met at sites or drainage areas that have less than 15% impervious cover, while drainage areas having no impervious cover and no proposed disturbance during development may be excluded from the WQ_v calculations. While the WQ_v is the storage volume needed to capture and treat the runoff from 90% of the average annual rainfall, it also provides management at a critical level (one-third bankfull elevation) within stream channels

Recharge Volume (Re_v)

The criteria for maintaining recharge is based on the average annual recharge rate of the hydrologic soil group(s) present at a site as determined by the United States Department of Agriculture Natural Resources Conservation Service (USDA-NRCS) soil surveys or from detailed soil investigations. Calculation of the specific recharge factor (S) for each soil is based on the USDA-NRCS average annual recharge volume per soil type divided by the annual rainfall (42 in. in the case of Maryland) and multiplied by 90%, consistent with the WQ_v methodology. The recharge volume is considered part of the total WQ_v that must be addressed at a site and can be achieved either by nonstructural techniques (e.g., buffers, runoff disconnection), structural practices (e.g., infiltration, bioretention), or a combination of both. Re_v and WQ_v are inclusive in that drainage areas having no impervious cover and proposed disturbance may be excluded from recharge calculations as well. The intent of the recharge requirement is to maintain existing groundwater recharge at development sites to preserve water table elevations in order to maintain the hydrology of streams and wetlands under dry-weather conditions. Re_v is dependent on slope, soil type, vegetative cover, precipitation, and evapotranspiration; sites with natural ground cover exhibit higher recharge rates when compared to impervious surfaces resulting from development.

Channel Protection Volume (Cp_v)

The channel protection storage volume(Cp_v) requirement exists to protect stream channels from excessive erosion caused by the increase in runoff from new development. The rationale for this criterion is that runoff from the 1-yr design storm will be stored and released in such a gradual manner that critical erosive velocities during bankfull and near-bankfull events will rarely be exceeded in downstream channels. The Cp_v requirement does not apply to direct discharges to tidal waters.

Of these criteria, the water quality, recharge and channel protection volumes are determined by soils, amount of imperviousness, proposed design and/or layout, and implementation of nonstructural practices. This simplifies calculations, reduces error and/or abuse, and provides direct incentives to reduce impervious areas. Another important feature of these three volumetric criteria is the relation to natural hydrologic processes. When considered together, these three criteria capture and treat the runoff from at least 95% of the average annual rainfall and mimic natural recharge and channel forming processes.

There are two primary approaches for managing stormwater runoff and addressing the unified stormwater sizing criteria requirements on a development site:

- the use of better site design practices to reduce the amount of stormwater runoff and pollutants generated and/or provide for natural treatment and control of runoff, and
- the use of structural stormwater controls to provide treatment and control of stormwater runoff (Atlanta Regional Commission, 2001).

Structural stormwater controls should be considered after all reasonable attempts have been made to minimize stormwater runoff and maximize its control and treatment through better site design methods. Once the need for structural controls has been established, all relevant stormwater sizing criteria should be applied in selecting one or more appropriate controls to meet the stormwater runoff storage and treatment requirements. Most development sites generally require a combination of structural and/or nonstructural BMPs to meet all stormwater sizing criteria (WEF and ASCE, 1998).

3.2.2 BMP Performance Objectives

A fundamental objective of stormwater management programs should be to attempt to reduce the change from the pre-development hydrology of the site. The use of structural controls for treating stormwater and to improve the quality of receiving waters has a set of objectives derived from a number of sources that include: (i) federal, state and local regulatory requirements; (ii) state or local community goals to mitigate the impacts associated with urban runoff; and, (iii) special local area needs such as trout or salmon fisheries protection, water supply and watershed protection, flood control to protect human life and property, groundwater protection, and other issues of local importance. In general, five different levels of stormwater BMP performance goals have been identified (Clar *et al.*, 2003).

- Level 1 - Flood control and peak discharge control
- Level 2 - Level 1 + 80% TSS removal
- Level 3 - Flood, peak discharge, and water quality control
- Level 4 - Unified sizing (multi-parameter) criteria
- Level 5 - Ecologically sensitive stormwater management

These goals have been discussed in detail in an earlier U.S. EPA publication on structural BMP design considerations for improving water quality(Clar *et al.*, 2003).

3.3 Design of Structural BMPs to Improve Water Quality

Stormwater BMPs can be designed for a wide range of goals and objectives, e.g., from a single parameter approach such as flood control typical in older developed watersheds, or pollutant removal typical of undeveloped watersheds , to multi-parameter ecological sustainability of receiving systems. These management goals will determine the requirement for the proper design and mix of ecological and engineering factors that must be considered. These typically include hydrology and inflow hydraulics, soil characteristics/infiltration rates, site-specific water quality concerns, location and site constraints, and the associated costs as well as the condition of the receiving waters (Clar *et al.*, 2003).

This book emphasizes that a clear understanding of the fundamental mechanisms at play within a BMP to reduce the effluent load is the key to properly designing BMPs. The preliminary step is the examination of the following criteria: (i) how does the BMP address the twin issues of stormwater quality and quantity control; (ii) what are the design considerations involved in achieving optimum water quality control in the BMP; (iii) how effective is the BMP in terms of meeting performance objectives; (iv) what are the cost factors associated with the design and construction of the BMP and how to be cost-effective; and more importantly, (v) how best to place it in the watershed. This book focuses on the design considerations of the following more commonly used structural BMPs:

3.3.1 Dry Extended-detention Ponds
3.3.2 Wet Ponds
3.3.3 Stormwater Wetland
3.3.4 Grassed Swales
3.3.5 Vegetated Filter Strips
3.3.6 Infiltration Trenches
3.3.7 Porous Pavements
3.3.8 Sand and Organic Filters

The following sections present a detailed discussion on the general design considerations and BMP design guidelines for these BMPs adopted by various state agencies for stormwater management. Several well known BMP manuals, (e.g.,Atlanta Regional Commission, 2001; Schueler, 1987;

Florida DER, 1988; NVPDC and ESI, 1992; Schueler, 1992a; WEF and ASCE, 1998; VA DCR, 1999; MDE, 2000; Metro Council, 2001; NYSDEC, 2001; U. S. EPA, 2001; WA DOE, 2001; Caltrans, 2002; GeoSyntec and ASCE, 2002; CASQA, 2003) have been consulted in presenting the multitude of approaches practiced by the different agencies in designing structural BMPs for stormwater management for new and existing urban development. It must be noted that the ensuing discussion is merely a compilation of the existing design practices and should not be treated as the ultimate design guidance document. The actual design process must take into account several locally specific criteria that primarily include the long-term rainfall/hydrological and soil considerations, water quality objectives, economics, and stormwater regulatory considerations. The typical operation and maintenance requirements for these BMPs are discussed in chapter 5.

3.3.1 Dry Extended-detention (ED) Ponds

An ED pond is an impoundment that temporarily stores stormwater runoff from a water quality design storm for a specified minimum period of time (usually 24 to 48 h) and discharges it through a hydraulic outlet structure to a downstream conveyance system; it is usually dry during non-rainfall periods and does not have any permanent standing water (CASQA, 2003; VA DCR, 1999). An ED pond can be designed to provide for one or all of the following: water quality enhancement; downstream flood control; and, channel erosion control. Conventional ED ponds temporarily detain a portion of stormwater runoff for up to 24 h after a storm using a fixed orifice resulting in the settling out of urban pollutants; enhanced ED ponds are designed to prevent clogging and suspension encountered in conventional ED ponds due to frequent high inflow velocities, and thus have higher efficiencies. ED ponds provide greater flexibility in achieving target detention times. Along with a detention area, they typically include a sediment forebay near the inlet, a micropool and/or plunge pool at the outlet, and utilize an adjustable reverse-sloped pipe as the ED pond control device to prevent resuspension of particles deposited in earlier storms (WY DEQ, 1999). The detention of runoff for 24 h or more (up to 72 h) helps to remove up to 90% of particulate pollutants, while the removal of soluble forms of nitrogen and phosphorus in urban runoff can be enhanced if the normally inundated area of the pond is managed as a shallow marsh or a permanent pool (Schueler, 1987).

3.3.1.1 Stormwater Control

ED ponds can be designed for flood control by providing additional storage above the ED volume, and by reducing the peak rate of runoff from the drainage area (CASQA, 2003). These BMPs are effective in controlling post-development peak discharge rates to the desired pre-development levels for the design storm(s) specified, and the optimum level is achieved by controlling multiple design storms. The design storms are chosen based on ordinance, or specified watershed conditions (VA DCR, 1999). ED ponds are also capable of managing smaller floods that contribute to channel erosion problems and occur more frequently than the annual or 2-yr flood (Schueler, 1987).

3.3.1.2 Pollutant Removal Capability

Pollutant removal is primarily accomplished by gravitational settling that is dependent on the detention time and the fraction of the annual runoff volume that is effectively detained in the pond (Schueler, 1987) together with sorption of pollutants to particulates and the associated settling velocity distribution of these particles. Conventional ED ponds provide moderate but variable removal of particulate pollutants such as sediment, phosphorus and organic carbon, and some removal of soluble pollutants (CASQA, 2003). Increasing detention times may result in greater removal of soluble pollutants. Urban pollutants commonly of concern, including nitrate and orthophosphates, remain in solution, and may be removed by managing the lower stage of the ED pond as a shallow wetland that utilize natural biological removal processes (Schueler, 1987; VA DCR, 1999).

Positive factors influencing pollutant removal (WY DEQ, 1999) include:

- six to twelve hours of minimum detention,
- smaller treatment volumes (e.g., 0.5 watershed in.) provide the best removal rates,
- wetlands in lower stage of design prevent resuspension and augment removal of sediments,
- use of a micropool to protect the ED pond orifice, and
- soils that soak up runoff and evapotranspire the same.

Negative factors influencing pollutant removal (WY DEQ, 1999) include:

- re-suspension of previously deposited pollutants from the pilot channel of pond floor,
- large treatment volumes (acceptable ED times cannot be achieved over the broad range of expected storms), and
- difficulty in predicting ED hydraulics.

3.3.1.3 Design Considerations

General Considerations

The success of an ED pond is dependent on the designer's ability to identify any site and downstream conditions that may affect the design and function of the pond. The facility should be compatible with both upstream and downstream stormwater systems to promote a watershed approach in providing stormwater management (VA DCR, 1999). The size can be based on the volume for which BMP credit is desired and the volume being dictated by stormwater management requirements, as these facilities are usually designed for both water quality and stormwater management needs. The shape of these facilities is often dictated by site constraints and topography (WEF and ASCE, 1998).

The dimensions of the pond need to be sized appropriately in order to enhance the effectiveness of these BMPs. An effective configuration of the pond should result in a long flow path, promote the establishment of low velocities, and avoid having stagnant areas of the pond. In order to promote settling of pollutants and aesthetic appeal, the design should consider the length-to-width ratio, cross sectional areas, pond slopes and configuration, and aesthetics.

Sizing Detention Ponds

There are several ways to size an ED pond. The more common methods use either maximized volume or hydrograph routing (WEF and ASCE, 1998).

The maximized volume method is the simplest and most direct way for smaller catchments serving up to approximately 1 km^2 (0.6 mi^2). The methodology to estimate the maximized water quality capture volume is described below.

The stormwater quality capture volume may be found by using continuous hydrologic simulation and local long-term hourly (or lesser increment) precipitation records, or by obtaining a first-order estimate of the needed capture volume using simplified procedures that target the most typically occurring population of runoff events. In the U.S., this data is available for ponds that empty their entire volume in 24 and 48 h. Mean values for other emptying times can be determined by interpolating the results for the 24 and 48 h time. After an extensive analysis of the mean annual runoff-producing rainfall depths for the different meteorological regions of the U.S., simple regression equations were established to relate the mean precipitation depth to "maximized" water quality runoff capture volumes (WEF and ASCE, 1998). The analytical procedure was based on a simple transformation of each storm's volume of precipitation to a runoff volume using a coefficient of runoff. A third order regression equation, Equation 3-1 (WEF and ASCE, 1998), was derived using data from more than 60 urban watersheds (U.S. EPA, 1983). The equation has broad applicability for smaller storm events in the U.S. as it was derived from a nationwide monitoring over a 2-yr period.

$$C = 0.858i^3 - 0.78i^2 + 0.774i + 0.04 \qquad \textbf{(3-1)}$$

where
C = runoff coefficient; and
i = watershed imperviousness ratio, namely, percent total imperviousness divided by 100.

Equation 3-2 relates mean precipitation depth to the "maximized" detention volume and is given by

$$P_0 = (a \times C) \times P_6 \qquad \textbf{(3-2)}$$

where
P_0 = maximized detention volume determined using either the event capture volume or the volume capture ratio as its basis, watershed in. (mm);
a = regression constant from least-squares analysis;
C = watershed runoff coefficient; and
P_6 = mean storm precipitation volume, watershed in. (mm).

Values of coefficient a have been determined based on an analysis of long-term data from seven precipitation gauging sites located in different meteorological regions of the U.S. for different drain times of 12, 24 and 48 h, respectively and shown in Table 3-2 (Guo and Urbonas, 1995). The correlation of determination coefficient, r^2, ranges from 0.80 to 0.97, implying a strong level of reliability. It is suggested that the event-capture-ratio-based coefficients in Table 3-2 be used with equation 3.2 instead of the volume capture ratio coefficients (WEF and ASCE, 1998). While the choice of the emptying or drain time rests with the designer or with the local authorities, it must be noted that suspended solids are better removed under longer emptying times. However, the disadvantage of longer drain times is that they tend to produce less attractive facilities, ones that have little or no vegetation on the bottom, or "boggy" bottoms with marshy vegetation that pose maintenance concerns.

Table 3-2. Values of Coefficient "*a*" for Finding the Maximized Detention Storage Volume*

		Drain time of capture volume		
		12 h	**24 h**	**48 h**
Event capture ratio	a r^2	1.109 0.97	1.299 0.91	1.545 0.85
Volume capture ratio	a r^2	1.312 0.80	1.582 0.93	1.963 0.85

* Approximately 85th percentile runoff event (range 82 to 88%)
(Guo and Urbonas, 1995)

The hydrograph routing method is used for detention ponds that serve areas larger than 1 km^2 by converting the maximized storm depth to a design hyetograph, to simulate a runoff hydrograph. Although the method by which this is done is dictated by the typical design storm temporal distribution in use within the region where the facility is located, it is suggested that the maximized depth be redistributed into a 2 h design storm hyetograph. The goal of reservoir routing is to balance inflow rates against outflow rates to find the needed volume, which can be accomplished with numerical methods or by using of one of the many available computer programs written for this purpose (WEF and ASCE, 1998). The needed storage volume is a time integral of the difference between inflow and outflow hydrographs from the beginning of storm runoff to the point in time where the outflow rate exceeds the inflow rate (Equation 3-3):

$$V_{\max} = \int_0^1 (Q_{in} - Q_{out})dt \tag{3-3}$$

where
V_{max} = storage volume;
t = time from beginning of runoff to a point of maximum storage;
$Q_{in} = Q_{out}$ on hydrograph recession limb;
Q_{in} = inflow rate; and
Q_{out} = outflow rate.

Local governments have developed a number of sizing rules for extended-detention, each specifying both a volume to be detained and a duration over which this volume is released. (CASQA, 2003) states that capture volume is determined by local requirements or sized to treat 85% of the annual runoff volume. It must be mentioned that a thorough engineering approach to sizing a pond may be the best option. This can be accomplished by using long-term dry- and wet-weather flows, pond inflow based on watershed hydrology, and particle settling velocity calculations together with necessary calculations for pond soil infiltration and evaporation. A pre-monitoring program to study the settling velocity distribution and the analysis of pollutant associations with particulates and soluble fraction is recommended.

Siting

Dry ED ponds are among the most widely applicable stormwater management practices and are especially useful in retrofit situations where their low hydraulic head requirements allow them to be sited within the constraints of the existing drainage system (CASQA, 2003). The basic guidelines for siting dry ED ponds are as follows:

- Dry ED ponds may be used for a wide range of drainage areas. However, the upper range for contributing drainage area applicable for these ponds without having to take baseflow into consideration is about 50 to 75 acres (NVPDC and ESI, 1992).

- Dry ED ponds should be used on sites with a minimum area of 5 acres. With this size catchment area, the orifice can be on the order of 0.5 in. The challenge in smaller sites is to provide channel or water quality control because the orifice diameter at the outlet needed to control relatively small storms becomes very small and is prone to clogging.

- The base of the extended-detention facility should not intersect the water table, as a permanently wet bottom may become a mosquito breeding ground (CASQA, 2003).

- Adequate access from a public or private right-of-way to the pond should be reserved. The access should be at least 10 ft wide, on a slope of 5:1 or less, and stabilized to withstand the passage of heavy equipment.

- All ED ponds should be a minimum of 20 ft from any structure or property line, and 100 ft from any septic tank/drain field. ED ponds should also be a minimum of 50 ft from any steep slope (greater than 15%). Otherwise, a geotechnical report will be required to address the potential impact of any pond that must be constructed on or near such a slope (VA DCR, 1999).

- ED ponds can be used with almost all soils and geology, with minor design adjustments for regions of rapidly percolating soils such as sand. In these areas, these BMPs may need an impermeable layer to prevent groundwater contamination. Highly permeable soils are not suitable for ED ponds, and for an enhanced ED pond, the soils must support the shallow marsh at the time of stabilization and planting.

Quantity Detained

The amount of runoff detained heavily influences the pollutant removal performance. At a minimum, ED ponds should be sized to accommodate the runoff produced by the mean storm, and preferably should be capable of storing the runoff volume of a 1.0 in. storm. Higher levels of control can be achieved when the runoff volume from the 1- or 2- yr storm is detained (Schueler, 1987). However, in many cases, the stricter storage requirements recommended above for streambank erosion control (1.0 to 1.5 in. R_v) will govern how much extra detention storage is needed.

Duration

Detention times of at least 24 to 36 h are probably necessary to achieve maximum removal of most pollutants. Although most of the settling occurs within the first 12 h in settling column experiments, it is advisable to provide further detention since several hours may be needed before ideal settling conditions develop in a pond. Slightly longer detention times of up to 40 h may be needed in larger watersheds for downstream channel erosion control. The control device must be sized so as to provide an adequate detention time for the entire spectrum of storms. The pond designer should perform several storage routing calculations e.g., TR-20 method or equivalent, to determine the approximate detention time for the smaller, more frequent runoff events. As a general rule, it is recommended that the average detention time for small runoff events (0.1 to 0.2 in.) should be no less than 6 h. As a final check, the runoff velocity of the downstream channel at the extended-detention release rate should be computed to make sure that it is not erosive (Schueler, 1987).

Pond Configuration

Minimizing the velocity of the flow through the pond greatly improves the pollutant removal efficiency of the pond, which can be effected by increasing the pond depth as well as the cross sectional area (NVPDC and ESI, 1992). The basin should gradually expand from the inlet and contract toward the outlet to reduce short circuiting and slow influent velocities by increasing the cross sectional flow area. The goal is to provide conditions where the velocity of flow through the facility for a typical storm event is less than the settling velocities of the pollutants of concern (NVPDC, 1979).

The length-to-width ratio of a pond is one design aspect that can significantly affect pollutant removal. The distance between inlet and outlet points needs to be maximized in order to promote pollutant settling. A high aspect ratio may improve the performance of detention ponds; consequently, the outlets should be placed to maximize the flowpath through the facility. While the flow path length is defined as the distance from the inlet to the outlet as measured at the surface, the average width is calculated as the surface area of the pond divided by the length (NVPDC and ESI, 1992) (Metro Council, 2001). A length-to-width ratio of two or greater, preferably up to a ratio of four is required for additional detention time for settling and biological treatment (WEF and ASCE, 1998).

Pond depths optimally range from 2 to 5 ft and may include a sediment forebay to provide the opportunity for larger particles to settle out (NVPDC and ESI, 1992).

A micropool is not recommended in the design because of vector concerns. For online facilities, the principal and emergency spillways must be sized to provide 10 ft of freeboard during the 25-yr event and to safely pass the flow from 100-yr storm (CASQA, 2003).

Pond Side Slopes

Pond side slopes need to be stable under saturated soil conditions. They also need to be sufficiently gentle to limit rill erosion, facilitate maintenance, and address the safety issue of individuals falling in when the basin is full of water. In order to promote facility effectiveness, it is highly desirable to avoid resuspension of materials collected on the pond floor; the potential for resuspension is generally minimized by reducing inflow velocities and maintaining vegetative cover. Side slopes should be no steeper than 4:1 (H:V), and no flatter than 20:1 (H:V) (Schueler, 1987; WEF and ASCE, 1998). Slopes steeper than this needs to be stabilized with an appropriate slope stabilization practice (CASQA, 2003).

Pond Lining

Ponds must be designed to prevent possible contamination of groundwater below the facility.

Pond Inlet

An ideal inflow structure should convey stormwater to the pond while preventing erosion of the pond bottom and banks, reducing resuspension of previously deposited sediment, and facilitating deposition of the heaviest sediment near the inlet. Such energy dissipation measures also reduce the tendency for short-circuiting. Inflow structures can be drop manholes, rundown chutes with an energy dissipator near the bottom, a baffle chute, a pipe with an impact basin, or one of the many other types of diffusing devices, depending on pond geometry (NVPDC and ESI, 1992; WEF and ASCE, 1998).

Outflow Structure

The outlet should be capable of slowly releasing the design capture volume over the design emptying time. ED ponds are designed to encourage sediment deposition and as stormwater has substantial quantities of settleable and floatable solids, outlets are prone to be clogged, invalidating the hydraulic function of even the best design. Each outlet therefore needs to be designed with clogging, vandalism, maintenance, aesthetics, and safety in mind (WEF and ASCE, 1998). The facility's drawdown time should be regulated by a gate valve or orifice plate. In general, the outflow structure should have a trash rack or other acceptable means of preventing clogging at the entrance to the outflow pipes. The structure should be sized in such a way to allow for complete drawdown of the WQ_v in 72 h, with no more than 50% of the water quality volume draining from the facility within the first 24 h. The outflow structure should be fitted with a valve in order to regulate the rate of discharge from the basin as well as to halt the discharge in case of an accidental spill in the watershed (CASQA, 2003). The discharge from a control orifice is given by Equation 3-4:

$$Q = CA(2gH - H_0)^{0.5} \qquad \textbf{(3-4)}$$

where
Q = discharge (ft^3/s);
C = orifice coefficient;
A = area of the orifice (ft^2);
g = gravitational constant (32.2 ft/s^2);
H = water surface elevation (ft); and
H_0 = orifice elevation (ft).

Recommended values for C are 0.66 for thin materials and 0.80 when the material is thicker than the orifice diameter. This equation can be used with the pond stage/volume relationship to calculate drain time in spreadsheet form.

Storage Volume

The storage volume in the ED pond should be equal to the maximized volume discussed earlier. An additional 20% could be added to this volume to provide for sediment accumulation, and could be used to promote sedimentation of smaller particles (less than 60 μm in size), which account for approximately 80% of the suspended sediment mass found in stormwater (WEF and ASCE, 1998)

Flood Control Storage

Whenever feasible, the ED pond should be incorporated within a larger flood control facility. By doing so, both water quality and flood control functions can be combined in a single detention basin.

Two-Stage Design

This pond configuration is meant to address both water quality and quantity. A two-stage basin is recommended when extended-detention is applied to dry ponds. The top stage of the pond should have the capacity to regulate peak flow rates of large infrequent storms (10-, 25- or 100- yr), and will generally remain dry between storms. The volume in this stage is called the "flood storage volume." The second stage of the pond is designed to detain smaller storms for a sufficient period of time to remove pollutants from the runoff. The volume in this stage is called the "water quality volume." For ED ponds, WQ_v is typically the runoff from the 0.3-yr storm event, since a large fraction of the annual pollutant load is delivered by small, frequent storm events (Metro Council, 2001).

The upper stage of the pond is sized and graded (2% minimum) to remain dry except during large infrequent storms, while the bottom stage is expected to be regularly inundated. The lower portion has a micropool that fills frequently and reduces the periods of standing water and sediment deposition in the remainder of the basin. A marsh-like environment in the lower section allows for some biological uptake of soluble materials and provides quiescent conditions, which promotes sedimentation of particulates (NVPDC and ESI, 1992). However, these recommendations do not necessarily apply to large, regional extended-detention ponds (WEF and ASCE, 1998). Extra storage, over and above stormwater and extended-detention requirements, should be provided within the bottom stage, or at the inlet to account for 20 yr of sediment deposition. The main advantage of this configuration is that frequently inundated areas are localized in one section of the pond, thus allowing the upper portion of the BMP facility to be used for certain low intensity recreational uses during dry weather.

Enhancement Options - Wetland Creation

Establishing wetland vegetation in a shallow marsh component or on an aquatic bench in the lower stage of the detention basin will enhance removal of soluble nutrients, increase sediment trapping, prevent sediment resuspension, and provide wildlife and waterfowl habitat (Metro Council, 2001). The use of a shallow marsh limits the maximum range of vertical storage in the ED pond to 3 ft above the marsh's water surface elevation. However, the surface area requirements for the shallow marsh will likely force the basin's geometry to broaden at the lower stages, which will compensate for the vertical storage (VA DCR, 1999). In general, extended-detention water surface elevations greater than 3 ft, and the frequency at which those elevations can be expected, are not conducive to the growth of dense or diverse stands of emergent wetland plants. Water depths of 6 to 12 in. would be required for optimal wetland growth, and native species should be planted in the wetland (Schueler, 1987). The general guidelines in designing a shallow marsh are discussed in detail in section 3.3.3 on stormwater wetlands.

Sediment Forebay

The settling area for incoming sediments can be increased through the addition of a sediment forebay. The use of a sediment forebay, however, is only recommended for wet ponds larger than 4,000 ft^3. The forebay is an excavated settling basin or a section separated by a low weir at the head of the primary impoundment. Forebays serve to trap sediments before the sediment enters the primary pool, effectively enhancing removal rates and minimizing long-term operation and maintenance problems. Also, it is easier and more cost effective to remove sediments from the forebay periodically as compared to removal from a wet pond pool. Hard bottom forebays make sediment removal easier, and forebays should be accessible to heavy machinery, if necessary. About 10 to 25% of the surface area of the wet pond should be devoted to the forebay.

The forebay can be distinguished from the remainder of the pond by one of several means: a lateral sill with rooted wetland vegetation; two ponds in series; differential pool depth; rock-filled gabions or retaining walls; or, a horizontal rock wall filter placed laterally across the pond. Energy dissipation techniques should be used at the inlet to the sediment forebay to avoid erosion, promote settling, and minimize short-circuiting of flows. The length-to-width ratio of the forebay should be at least 2:1 to minimize short-circuiting (Metro Council, 2001).

Low Flow Channels

Low flow channels route the last remaining runoff, dry weather flow and groundwater to the permanent pool and outlet. These channels should be installed in the upper stage of the basin to ensure that the basin dries out completely. Low flow channels also serve to prevent erosion of the upper stage of the pond outside as runoff first enters the pond.

The presence of a baseflow makes the design of an extended-detention control structure difficult. An orifice designed for wet-weather baseflow, compromises the dry-weather control due to very high release rates. On the contrary, if an orifice is undersized to meet the dry-weather control, the ED pond may remain full of water during the wet-weather season and eliminate the extended-detention volume by creating an undersized permanent pool (VA DCR, 1999). When seasonal baseflow is present, an adjustable orifice should be provided in the control structure to maintain the marsh volume.

Design considerations should take into account the presence of a baseflow and the associated potential for erosion within the basin and, ideally, spread them out so they sheet flow across the bottom of the basin. A few local ordinances require the use of low-flow channels to carry baseflows. Generally, an impervious low-flow channel is not recommended in a stormwater management water quality basin, as its use is contrary to the basin's water quality function. However, an impervious ditch may be used to carry baseflow if it is designed to overflow during storm events and spread the runoff across the basin floor. The use of gabion baskets or riprap, instead of concrete, may provide the advantage of slowing the flow, encouraging spillover onto the basin floor.

Overflow

Similar to a constructed stormwater wetland, an ED overflow system should be designed to provide adequate overflow or bypass for a full range of design storms.

Liner to Prevent Infiltration

ED ponds should have negligible infiltration rates through the bottom of the pond. If infiltration is anticipated, and the area is not suspected to be underlain by karst, then an infiltration facility, rather than a detention water quality BMP should be used or a liner should be installed in the basin to prevent infiltration. The following recommendations apply when using a liner:

- A clay liner should have a minimum thickness of 12 in. and should comply with the specifications provided in Table 3-3.
- A layer of compacted soil (minimum 6 to 12 in. thick) should be placed over the liner before seeding with an appropriate seeding mixture.
- Other liner types may be used if supporting documentation is provided verifying the liner material's performance.

Table 3-3. Clay Liner Specifications

Property	Test Method (or equal)	Unit	Specification
Permeability	ASTM D-2434	cm/sec	1×10^{-6}
Plasticity Index of Clay	ASTM D-423 & D-424	%	Not less than 15
Liquid Limit of Clay	ASTM D-2216	%	Not less than 30
Clay Particle Sieving	ASTM D-422	%	Not less than 30
Clay Compaction	ASTM D-2216	%	95% of Standard Proctor Density

(City of Austin, 1988)

Pond Buffer

A buffer strip away from the pond to the nearest lot should be reserved and landscaped using low-maintenance grasses, shrubs and trees (e.g., the minimum width should be 25 ft (Schueler, 1987). A landscaping plan for the pond and buffer should outline measures to improve the appearance for adjacent residents, meet specific design functions, and provide local wildlife habitat (Schueler, 1987).

Dam Embankment

The dam embankment should be designed not to fail during storms larger than the water quality design storm. An emergency spillway could be provided, the design of which is governed by local regulations, and the embankment should have at least one foot of freeboard above the emergency spillway. The other approach is to design the embankment to withstand overtopping commensurate with embankment size, the volume of water that can be stored in it, and the potential of downstream damages or loss of life if the embankment fails. Embankments for small onsite basins should be protected from at least the 100-yr flood, while the larger facilities should be evaluated for the probable maximum flood. Embankment slopes should be no steeper than 3:1, preferably 4:1 or flatter. They also need to be planted with turf-forming grasses. Embankment soils should be compacted to 95% of their maximum density at optimum moisture, graded to allow access for heavy equipment, and mowed twice a year to prevent woody growth. At least 10 to 15% extra fill should be allowed on the embankment to account for possible subsidence.

Vegetation

Vegetation provides erosion control and enhances sediment entrapment in a pond. The pond can be planted with native grasses or with irrigated turf, depending on the local setting, pond design, and its intended other uses such as recreation. The maintenance of a healthy grass cover on the pond bottom is difficult due to sediment deposition, along with frequent and prolonged periods of inundation. Options for an alternative bottom liner include a marshy wetland bottom, bog, layer of gravel, riparian shrub, bare soil, low-weed species, or other type that can survive the conditions existing in the bottom of the pond.

Splitter Box

Splitter structures isolate WQ_v, when ponds are designed as offline facilities. The splitter box, or other flow diverting approach, should be designed to convey the large storm event (e.g., 25- yr storm) while providing adequate (e.g., at least 10 ft) freeboard along pond side slopes.

Erosion Protection at the Outfall

For online facilities, special consideration should be given to the facility's outfall location. Flared pipe end sections that discharge at or near the stream invert are preferred. The channel below the pond outfall should be modified to conform to natural dimensions, and lined with large stone riprap placed over filter cloth. Energy dissipation may be required to reduce flow velocities from the primary spillway to non-erosive velocities. An example schematic of a dry ED pond (MDE, 2000) is shown in Figure 3-1.

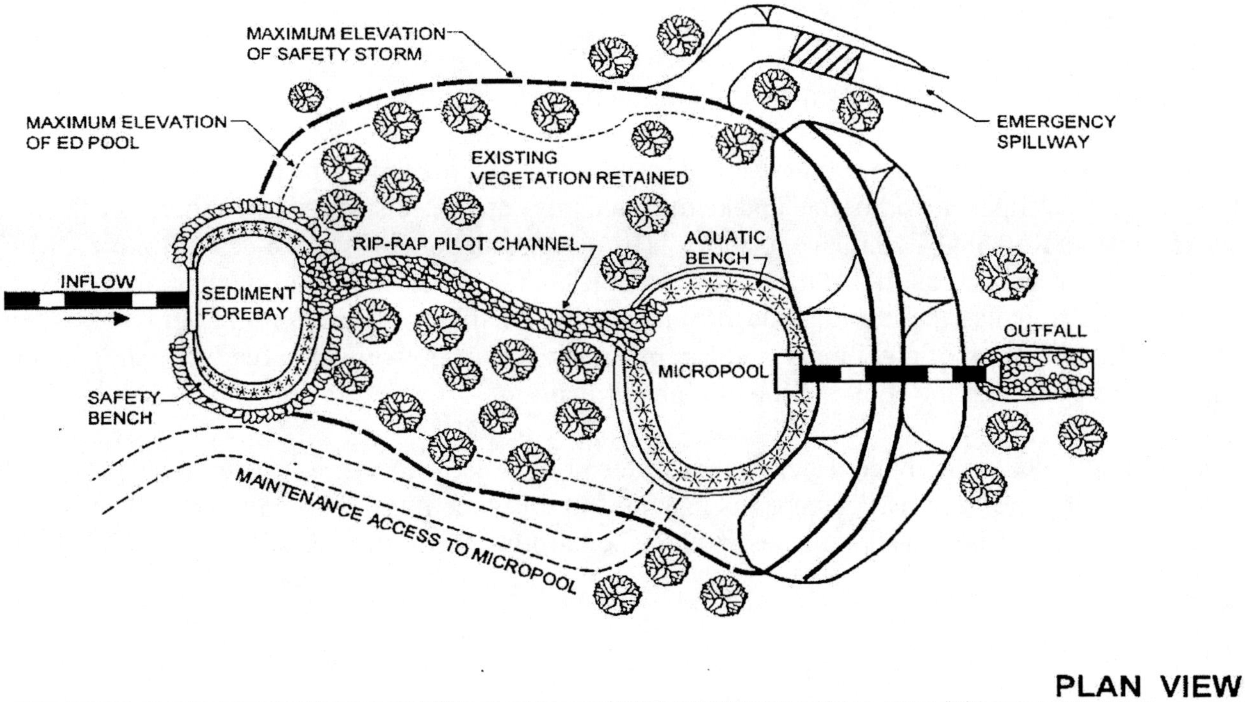

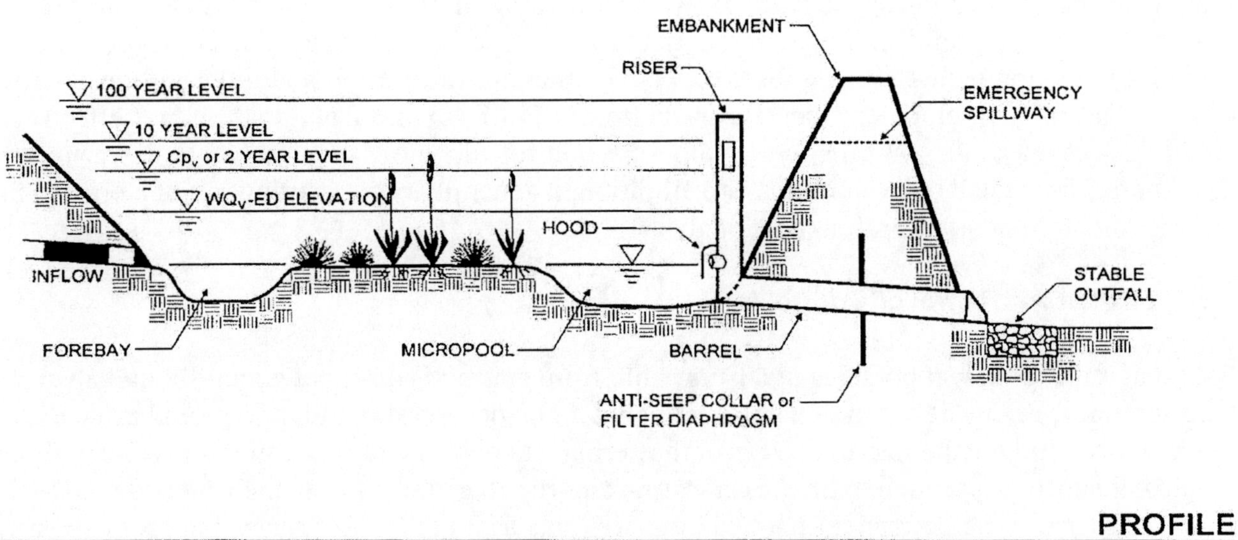

Figure 3-1. Schematic of a Dry Extended-Detention Pond (MDE, 2000)

3.3.2 Wet Ponds

A wet pond is a constructed stormwater retention basin with emergent wetland vegetation around the perimeter. It is designed to have a permanent pool of water throughout the year or at least during the wet season. Runoff from each rain event is detained and treated in the pool primarily through gravitational settling and biological uptake mechanisms until it is displaced by runoff from the next storm (Atlanta Regional Commission, 2001). The permanent pool provides a vessel for the settling of solids between storms and the removal of nutrients and dissolved pollutants. The wetland vegetation, also called the littoral zone, provides aquatic habitat, enhances pollutant removal, and reduces the formation of algal mats, and can be created by excavating an already existing natural depression or through the construction of embankments.

Conventional wet ponds have a permanent water pool to treat incoming stormwater runoff. In enhanced wet pond designs, a forebay is installed to trap incoming sediments where they can be easily removed; a fringe wetland is also established around the perimeter of the pond (WY DEQ, 1999). .

3.3.2.1 Stormwater Control

Stormwater ponds are designed to control both stormwater quality and quantity and can be used to address all the unified stormwater sizing criteria for a given drainage area. Wet ponds can be effective in controlling post-development peak discharge rates to pre-development levels for desired design storms. Groundwater recharge in wet ponds is limited to the storage lost to infiltration through the pond bottom. Although the quantity of recharge is greater than that achieved in dry or extended detention ponds, it is negligible in comparison to infiltration and other volume control BMPs. The post-development increase in the total runoff volume from a site is not effectively modified by wet ponds. While some temporary control of runoff volume happens when extra dead-storage is created by evaporation or infiltration, it generally occurs during minor storms in the summer months and after prolonged droughts.

3.3.2.2 Pollutant Removal Capability

Pollutant removal in wet ponds is highly variable from storm to storm but generally high over the long-term, for well designed and maintained ponds. The degree of pollutant removal achieved by a pond is a function of the size and design of the permanent pool and the characteristics of individual urban pollutants. Suspended sediments in stormwater runoff settle out from the water column to the pond sediments. The permanent pool additionally acts as a barrier to resuspension of deposited materials and improves removal performance over that achieved by dry ponds. The greatest initial settling often occurs near the pond inlet under quiescent conditions; settling can be modeled assuming Stokes Law Type I sedimentation. However, pollutant removal rates may decline during larger storms in smaller ponds due to short-circuiting and the volume of incoming runoff being greater than the volume of the permanent pool.

A unique feature of wet ponds is the presence of aquatic plants and algae that can remove significant amounts of soluble nutrients from the water column; retention ponds can be superior to ED ponds for the control of dissolved nutrients in stormwater (Schueler, 1987). Since soluble nutrients have minimal settling velocities, biological uptake represents an important removal pathway. Retention ponds are most appropriate where nutrient loadings are of concern, especially in the following situations:

- Watersheds tributary to reservoirs and lakes - retention ponds in the watershed can help achieve eutrophication management goals in downstream reservoirs and lakes.
- Watersheds tributary to tidal embayments and estuaries - nutrient loadings into estuarine systems is a growing concern in coastal areas, including upland areas that drain into tidal waters; retention ponds can help reduce the nutrient loads.

The degree of pollutant removal is a function of pool size in relation to contributing watershed area, and is achieved by gravitational settling, algal settling, wetland plant uptake, and bacterial decomposition. Unlike ED ponds, wet ponds avoid resuspension, and nutrient cycling in these ponds is generally thought to operate much as in natural lakes; consequently, the pollutant removal capabilities can be successfully predicted by applying a controlled lake eutrophication model. The principal factors governing nutrient cycling are the loading and the decay rates for phosphorus, hydraulic residence time, and mean depth (NVPDC and ESI, 1992).

The observed pollutant removal of a wet pond is highly dependent on two factors; i.e., the volume of the permanent pool relative to the amount of runoff from the typical event in the area and the quality of the baseflow that sustains the permanent pool. If the permanent pool is much larger than the volume of runoff from an average event then the primary process is the displacement of the permanent pool by the wet-weather flow (Caltrans, 2002). The discharge quality of wet ponds during dry- and wet-weather flows is not significantly different, resulting in a relatively constant discharge quality during storms that is the same as the concentrations observed in the pond during ambient (dry weather) conditions and so are better characterized by the average effluent concentration, rather than the "percent reduction" (CASQA, 2003). The dry- and wet-weather discharge quality is thus related to the quality of the baseflow that sustains the permanent pool and the transformations of those pollutants during their residence in the basin.

Positive factors influencing pollutant removal include:

- pretreatment by sediment forebay,
- permanent pool, 0.5 to 1.0 in. per impervious acre treated,
- fringe wetlands,
- shallow wetlands and/or extended detention may improve removal efficiencies, and
- high length-to-width ratios.

Negative factors influencing pollutant removal include::

- small pool size,
- fecal contribution from large waterfowl populations,
- short-circuiting and turbulence,
- sediment phosphorus release,
- extremely deep pool depths (greater than 10 ft), and

- snowmelt conditions and/or ice.

3.3.2.3 Design Considerations

A well designed stormwater pond consists of the following:
- permanent pool of water,
- overlying zone in which runoff control volumes are stored, and
- shallow littoral zone (aquatic bench) along the edge of the permanent pool that acts as a biological filter (WEF and ASCE, 1998).

Siting

Wet ponds are a widely applicable stormwater management practice and can be used over a broad range of storm frequencies and sizes, drainage areas, and land use types. They can be constructed on- or off-line (off-line is preferred) and can be sited at feasible locations along established drainage ways with consistent baseflow. Wet basin application is appropriate in the following settings:

- where there is a need to achieve a reasonably high level of dissolved contaminant removal and/or sediment capture,
- in small to medium-sized regional tributary areas with available open space and drainage areas greater than about 10 ha (25 acre),
- where baseflow rates or other channel flow sources are relatively consistent year-round, and
- in residential settings where aesthetic and wildlife habitat benefits can be appreciated and maintenance activities are likely to be consistently undertaken (CASQA, 2003).

Soil Permeability

Highly permeable soils may not be acceptable for retention ponds because of excessive drawdown during dry periods. Where permeable soils are encountered, exfiltration rates can be minimized by scarifying and compacting a 0.3 m (12 in.) layer of the bottom soil of the pond, incorporating clay to the soil, or providing an artificial liner. Excavating the permanent pool into the groundwater table can also ensure its permanency, but seasonal fluctuations in the groundwater table need to be taken into account (Schueler, 1987).

Design Criteria

The design of permanent pools for a wet pond employs two different methods.
- The solids-settling design method relies on the solids-settling theory and assumes that all pollutant removal is because of sedimentation.
- The lake eutrophication model design method provides for a level of eutrophication by accounting for the principal nutrient removal mechanisms (WEF and ASCE, 1998).

The solids-settling method is most appropriate for situations where the control of total suspended sediments and pollutants that attach themselves to the solids is the principal objective. The method relies on rainfall and runoff statistics, pond size, and settling velocity distributions of suspended solids to calculate total suspended sediment removal. This method assumes an approximate plug flow system in the retention pond with all pollutant removal resulting from sedimentation. Testing this model using data from nine retention ponds monitored during U.S. EPA's NURP showed that it predicted removal rates reasonably well (WEF and ASCE, 1998).

The lake eutrophication model assumes that a retention pond is a small eutrophic lake that can be represented by empirical models used to evaluate lake eutrophication effects. This method is used to size a retention pond to achieve a controlled rate of eutrophication and an associated removal rate for nutrients. Retention ponds that achieve nutrient removal also removes other pollutants, and typically it is not necessary for the design process to address constituents other than nutrients.

Like most input/output lake eutrophication models, this model is an empirical approach that treats the permanent pool as a completely mixed system and assumes that it is not necessary to consider the temporal variability associated with individual storm events. While the solids-settling model accounts for the temporal variability of individual storms, the lake eutrophication model is based on annual flows and loadings.

The model is applied in two parts:

$$K = \frac{0.56 \times Q_s}{F \times (Q_s + 13.3)} \qquad \textbf{(3-5)}$$

where
K_2 = second order decay rate m^3/mg x a;
Q_s = Z/T the mean overflow rate, m/a;
Z = mean pond depth, m;
T = average hydraulic retention time, yr; and
F = inflow (ortho P/total P) ratio.

$$R = 1.0t\frac{1.0 - \sqrt{1.0 + (1.0 + 4N)}}{2N} \qquad \textbf{(3-6)}$$

where
R = total P retention coefficient, (i.e. BMP efficiency);
$N = K_2 \times P_T \times T$; and
P_T = inflow total P, µg/L.

These two equations were developed from a database for 60 U. S. Army Corps of Engineers' reservoirs and were verified for 20 other reservoirs. When this model was applied to 20 other reservoirs, 10 NURP sites and 14 other retention pond systems and small lakes, the goodness-of-fit test yielded an $R^2 = 0.8$, indicating a good job of replicating monitored total P removal (WEF and ASCE, 1998).

Sizing

State and regional stormwater management regulations and guidelines often address design criteria for the permanent pool storage volume in terms of either the average hydraulic retention time, or minimum total suspended sediment removal rate. The size of the permanent pool in relation to the contributing watershed is perhaps the single largest factor influencing pollutant removal in wet ponds. A number of wet pond sizing rules that variously specify the minimum volume of the permanent pool have been proposed to optimize pollutant removal. There is no individual rule that can be recommended as applicable in all cases, and the choice in many cases rests with local stormwater management policy-makers. Sizing should take into account the objective that the pond should be sized to hold the permanent pool as well as the required water quality volume. An example (Schueler, 1987) is provided in Table 3-4. From the table, it can be seen that the choice of an appropriate pond sizing rule necessarily invites a trade-off between the degree of removal efficiency desired and the cost of achieving it.

Table 3-4. Summary of Wet Pond Sizing Rules

Sizing Rule	Sediment removed	Phosphorus removed	Extra storage	Extra cost
			(compared to 2 yr dry pond)	
RULE 1: 0.5 in. runoff per acre	60-90%	35-90%	35-200%	20-90%
RULE 2: 0.5 in. runoff per impervious acre	60%	35-40%	30%	20-25%
RULE 3: 0.1 to 0.8 in. depending on land use	55-80%	30-50%	30-70%	20-40%
RULE 4: 2.5 times the runoff of the mean storm	75%	55%	75%	40-50%
RULE 5: 4.0 times the runoff of the mean storm (app. 2 week retention)	85-90%	65%	200-250%	80-100%

(Schueler, 1987)

Pond Shape and Geometry

The wet basin should be configured as a two-stage facility with a sediment forebay and a main pool. Long, narrow and irregular shapes are also desirable for shallower ponds since they reduce surface area exposed to the wind and thereby prevent resuspension of previously deposited materials (Schueler, 1987). The basin should be wedge-shaped, narrowest at the inlet and widest at the outlet (CASQA, 2003).

The minimum length-to-width ratio for the permanent pool shape is 1.5:1, and should ideally be greater than 3:1 to avoid short-circuiting. Baffles, pond shaping or islands can be added within the permanent pool to increase the flow path (Atlanta Regional Commission, 2001).

Mean depth of the permanent pool is calculated by dividing the storage volume by the surface area. The mean depth should be shallow enough to ensure aerobic conditions and reduce the risk of thermal stratification but deep enough to ensure that algal blooms are not excessive and reduce resuspension of settled pollutants during significant storm events.

The minimum depth of the open water area should be greater than the depth of sunlight penetration to prevent emergent plant growth in this area, namely, on the order of 2 to 2.5 m (6 to 8 ft). A mean depth of approximately 1 to 3 m (3 to 10 ft) should produce a pond with sufficient surface area to promote algal photosynthesis and should maintain an acceptable environment within the permanent pool for the recommended average hydraulic retention times, although separate analyses should be performed for each locale (WEF and ASCE, 1998).
If the pond has more than 0.8 ha (2 acre) of water surface, mean depths of 2m (6.5 ft) will protect it against wind generated resuspension of sediments. A water depth of approximately 1.8 m (6 ft) over the major portion of the pond will also increase winter survival of fish (Schueler, 1987). A maximum depth of 3 to 4 m (10 to 13 ft) should reduce the risk of thermal stratification; however, in the state of Florida, pools up to 9.2 m (30 ft) deep have been successful when excavated in high groundwater areas, probably because of improved circulation at the bottom of the pond as a result of the movement of groundwater through it.

The perimeter of all permanent pool areas with depths of 4 ft or greater should be surrounded by an aquatic bench that extends inward 5 to 10 ft from the perimeter of the permanent pool and should be no more than 18 in. below normal depth. The area of the bench should not exceed about 25% of pond surface. The depth in the center of the basin should be 4 to 8 ft deep to prevent vegetation from encroaching on the pond open water surface.

Side Slopes Along Shoreline and Vegetation

Side slopes along the shoreline of the retention pond should be 4H:1V. CASQA (2003) recommends 3:1 or flatter to facilitate maintenance and reduce public risk of slipping and falling into the water. Additionally, a littoral zone should be established around the perimeter of the permanent pool to promote the growth of emergent vegetation along the shoreline and deer populations from wading.

This bench for emergent wetland vegetation should be at least 3 m (10 ft) wide with a water depth of 0.15 to 0.45 m (0.5 to 1.5 ft). The total area of the aquatic bench should be 25 to 50% of the permanent pool's water surface area. The use of wetland vegetation within shallow sections of the permanent pool should adhere to guidelines issued by local agricultural agencies or commercial nurseries. Emergent plants such as bulrush, three-square and lizards tail can provide an attractive fringe habitat (Schueler, 1987), providing food and cover for wildlife and waterfowl.

Extended-detention Zone Above the Permanent Pool

Some state or local regulations require detention of a specified runoff volume as surcharge above the permanent pool, in order to reduce short-circuiting and enhance settling of total suspended sediments. Although the addition of an extended-detention zone above the permanent pool may not likely produce measurable increases in the removal of total suspended sediments, it is still recommended to have a surcharge extended detention volume, and whenever one is used or required, it is suggested in these local guidelines that the maximum event-based volume with a 12 h drain time be used..

Minimum and Maximum Tributary Catchment Areas

Stormwater ponds should have a minimum contributing drainage area of 25 acres or more for a wet pond to maintain a permanent pool. The minimum drainage area should permit sufficient baseflow to prevent excessive retention times or severe drawdown of the permanent pool during dry seasons. It is recommended that a water balance calculation be performed using local runoff, evapotranspiration, exfiltration, and baseflow data to ensure that the baseflow is adequate to keep the pond full during the dry season. The maximum tributary catchment area should be set to reduce the exposure of upstream channels to erosive stormwater flows, reduce effects on perennial streams and wetlands, and reduce public safety hazards associated with dam height (WEF and ASCE, 1998).

Construction of Retention Ponds in Wetland Areas

Although wet pond BMPs are typically designed to enhance pollutant removal by incorporating wetland areas along the perimeter, regulatory agencies may restrict their use if a significant amount of native wetlands will be submerged within the permanent pool. If field inspections indicate that a significant wetlands area will be affected at a particular site, and if the construction of a wet pond is inevitably required, the following options may be pursued during final design subject to approval: investigate moving the embankment and permanent pool upstream of the major wetland area; and if this is not feasible, a wetland mitigation plan can be developed as a part of the retention pond design.

Forebay

A sediment forebay with a hardened bottom should be constructed near the inlet to trap coarse sediment particles in order to reduce the frequency of major clean-out activities within the pool area. The forebay storage capacity should be approximately 10% of the permanent pool storage and should be at least 3 ft deep (WEF and ASCE, 1998; CASQA, 2003). Exit velocities from the forebay should not be erosive. A fixed vertical sediment depth marker should be installed in the forebay to measure sediment accumulation. Access for mechanized equipment should be provided to facilitate removal of sediment. The forebay can be separated from the remainder of the permanent pool by one of several means: a lateral sill with wetland vegetation; two ponds in series; differential pool depth; rock-filled gabions; a retaining wall; or a horizontal rock filter placed laterally across the permanent pool.

Inlet and Outlet Structures

The inlet design should dissipate flow energy and diffuse the inflow plume where it enters the forebay or permanent pool. Examples of inlet designs include drop manholes, energy dissipaters at the bottom of paved rundowns, a lateral bench with wetland vegetation, and the placement of large rock deflectors.

An outlet for a retention pond typically consists of a riser with a hood or trash rack to prevent clogging and an adequate anti-vortex device for basins serving large drainage areas. A few examples are outlet works with surcharge detention for water quality, negatively sloped pipe outlet with riser, and multiple orifice outlet. Anti-seep collars should be installed along outlet conduits passing through or under the dam embankment. If the pond is part of a larger peak-shaving detention basin, the outlet should be designed for the desired flood control performance. An emergency spillway must be provided and designed using accepted engineering practices to protect the basin's embankment. The pond embankment and spillway should be designed in accordance with federal, state, and local dam safety criteria. For on-line facilities, the principal and emergency spillways must be sized to provide 1.0 ft. of freeboard during the 25- yr event and to safely pass the 100- yr flood (CASQA, 2003). The channel that receives the discharge from the basin's outlet should be protected from erosive discharge velocities. Options include riprap lining of the channel or providing stilling basins, check dams, rock deflectors, or other devices to reduce outfall discharge velocities to nonerosive levels.

When the pond is designed as an off-line facility, a splitter structure is used to isolate the water quality volume. The splitter box, or other flow diverting approach, should be designed to convey the 25- yr event while providing at least 1.0 ft of freeboard along pond side slopes.

Each pond must have a bottom drain pipe with an adjustable valve that can completely or partially drain the pond within 24 h. However, this requirement may be waived for coastal areas, where positive drainage is difficult to achieve due to very low relief (Atlanta Regional Commission, 2001). The pond drain pipe should be sized one pipe size greater than the calculated design diameter. The drain valve is typically a handwheel-activated knife or gate valve. Valve controls shall be located inside of the riser at a point where they (i) will not normally be inundated, and (ii) can be operated in a safe manner. Figure 3-2 provides an example schematic of a wet pond (CASQA, 2003)

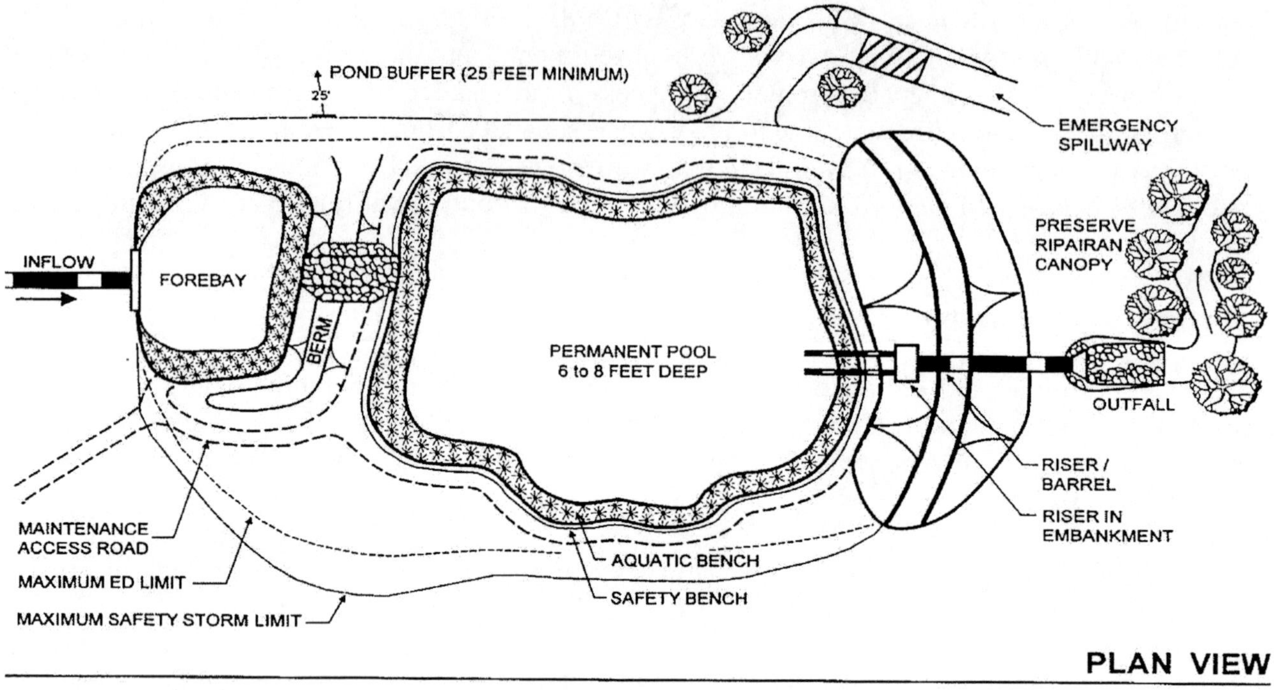

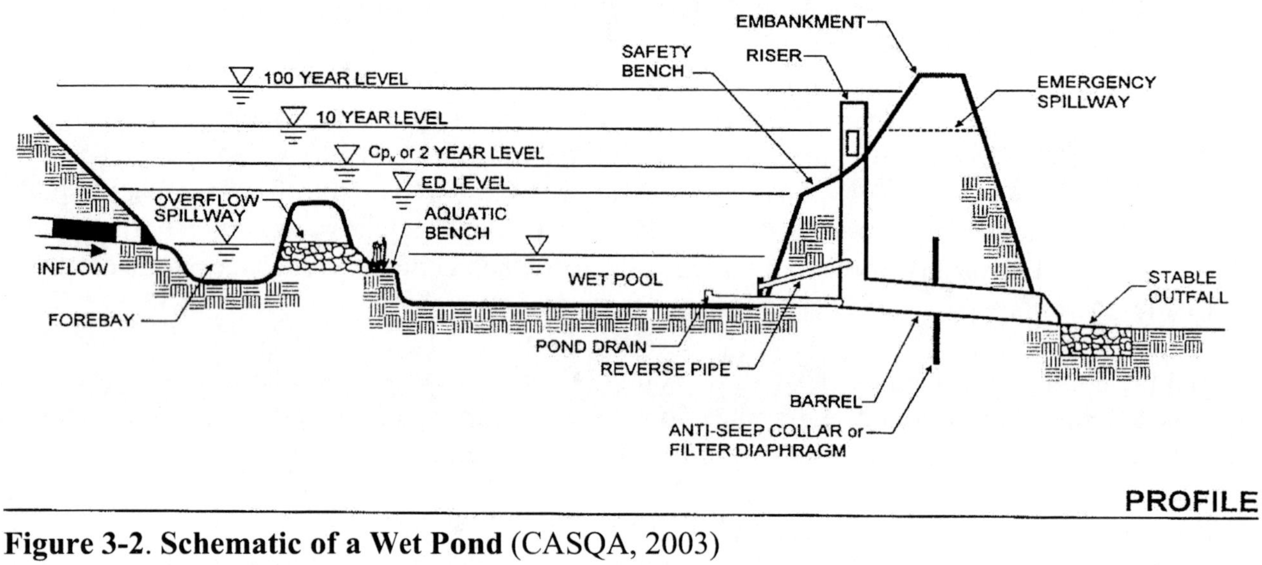

Figure 3-2. **Schematic of a Wet Pond** (CASQA, 2003)

3.3.3 Stormwater Wetlands

Stormwater wetlands can be defined as constructed wetland systems that are explicitly designed to mitigate the impacts of stormwater quality and quantity that occur during the process of urbanization. They do so by temporarily storing stormwater runoff in shallow pools that create growing conditions suitable for emergent and riparian wetland plants and routing runoff through vegetation to maximize contact. The runoff storage, complex micro topography and emergent plants in the stormwater wetland together form an ideal matrix for the removal of urban pollutants (Schueler, 1992a). Stormwater wetlands have been characterized as having one of five basic designs: (1) shallow marsh system; (2) pond/wetland system; (3) extended detention wetland; (4) pocket wetlands; and, (5) fringe wetlands.

Conventional stormwater wetlands are shallow pools that create growing conditions suitable for the growth of marsh plants. These are constructed systems and typically are not located within delineated natural wetlands. Stormwater wetlands differ from other artificial wetlands created to comply with mitigation requirements in that they may not replicate all the ecological functions of natural wetlands. However, as with natural wetlands, stormwater wetlands require a continuous baseflow or a higher table to support aquatic vegetation (Atlanta Regional Commission, 2001). Functional differences depend on the design of the wetland, interactions with groundwater and surface water, and local storm climate (WY DEQ, 1999). Enhanced stormwater wetlands designed for more effective pollutant removal and species diversity also include design elements such as a forebay, complex microtopography, and pondscaping with multiple species of wetland trees, shrubs and plants.

3.3.3.1 Stormwater Control

Constructed stormwater wetlands should generally not be used for flood control or stream channel erosion control due to the anticipated water level fluctuations associated with quantity controls (VA DCR, 1999). The clearing of vegetation and the addition of impervious surfaces may cause large and sudden surges of runoff during rain events, and may cause less than normal baseflows during dry periods. Large, sudden fluctuations in water levels can stress emergent wetland and upland edge vegetation, most of which cannot survive drought or saturation extremes, leaving wetland banks exposed to potential erosion. The large surface area requirement for constructed stormwater wetlands will help to minimize the "extreme" water level fluctuations during all but the larger storm events and the wetland design should allow for gradual increases and increases in wetland design.

3.3.3.2 Pollutant Removal Capability

Wetlands remove pollutants through gravitational settling, wetland plant uptake, adsorption, physical filtration and microbial decomposition. Primary removal of stormwater pollutants occurs during the relatively long quiescent period between storms (WY DEQ, 1999) and the degree of pollutant removal is a function of aquatic treatment volume, surface area-to-volume ratio, and the ratio of wetland surface area to watershed area. Pollutant removal is also expected to increase with longer storm water flow paths through the wetland and longer residence time within the wetland.

Conventional stormwater wetlands have a high pollutant removal capability that is generally comparable to that of wet ponds. While sediment removal may be greater in well designed facilities, phosphorus removal is more variable. Some cases of negative removal for ammonia and orthophosphorous are reported; the addition of ammonia or orthophosphorous may be due, in part, to wildlife use and populations and vegetation management (WY DEQ, 1999). According to Strecker *et al.*, (1990), overall performance is greatest during the growing season and lowest during the winter months.

Positive factors influencing pollutant removal include:

- constant pool elevations,
- range of micro-topography within the watershed,
- sediment forebay,
- high surface area to volume ratio,
- constructed wetland performs better than natural wetland,
- adding greater retention volume and/or detention time to the wetland,
- effective in areas with high water table or poorly drained soils, and
- lengthy travel paths for stormwater.

Negative factors influencing pollutant removal include::

- lower removal rate during non-growing seasons,
- concentrated inflows,
- sparse wetland cover, and
- ice cover or snowmelt runoff that would require a modification in design.

3.3.3.3 Design Considerations

General Considerations

A well-designed stormwater wetland consists of:

- shallow marsh areas of varying depths with wetland vegetation,
- permanent micropool, and
- overlying zone in which runoff control volumes are stored.

In addition, all wetland designs must include a sediment forebay at the inflow to the facility to allow heavier sediments to drop out of suspension before the runoff enters the wetland marsh.
Additional pond design features include an emergency spillway, maintenance access, safety bench, wetland buffer, and appropriate wetland vegetation and native landscaping (Atlanta Regional Commission, 2001).

Specific site conditions are important to the proper design of a wetland. Key site characteristics include soils, hydro period, and plant species and density. Depth to the confining layer or groundwater is important to ensure that the wetland does not dry up during extended periods of no rainfall. In addition, a constant source of surface water is recommended, taking appropriate measures to prevent the undesirable consequences of stagnant water in the wetlands. The depth and duration of maximum submergence are important because an excess of either will kill the vegetation (WEF and ASCE, 1998).

Location and Siting

A continuous baseflow or high water table is required to support wetland vegetation. A water balance must be performed to demonstrate that a stormwater wetland can withstand a 30-day drought at summer evaporation rates without completely drawing down (Atlanta Regional Commission, 2001).

Stormwater wetlands should normally have a minimum contributing drainage area of 25 acres or more and the minimum drainage area is 5 acres for a pocket wetland.

Wetland siting should also take into account the location and use of other site features such as natural depressions, buffers, and undisturbed natural areas, and should attempt to aesthetically fit the facility into the landscape. Bedrock close to the surface may prevent excavation.

Stormwater wetlands cannot be located within navigable waters of the U.S., (including wetlands), without obtaining a Section 404 permit under the Clean Water Act and any other applicable state permit. In some isolated cases, a wetlands permit may be granted to convert an existing degraded wetland through local watershed restoration efforts.

Minimum setback requirements for stormwater wetland facilities unless specified by local ordinances or criteria are as follows:

- from a property line - 10 ft,
- from a private well - 100 ft; if well is downgradient from a hotspot land use then the minimum setback is 250 ft, and
- from a septic system tank/leach field - 50 ft.

If a wetland facility is not used for overbank flood protection, it should be designed as an offline system to bypass higher flows rather than passing them through the wetland system.

Sizing

For optimal pollutant removal, a stormwater wetland must meet the following seven basic sizing criteria:

- contain a treatment volume (V_t) that is capable of capturing the runoff generated by 70 to 90% of the runoff-producing storms in the region on an annual basis,
- have a minimum surface area in relation to the contributing watershed area,
- allocate the surface area of the wetland to meet targets for certain depth zones,
- meet a minimum standard for the internal flow path through the wetland,
- demonstrate that the water supply to the wetland is greater than the expected loss rate so that water elevations can be maintained, and
- provide for extended detention for smaller storms (for ED wetlands only)(Schueler, 1992b).

Physical Specifications/Geometry

Recommended hydraulic design criteria for wetlands are as follows:

- Maintain dry-weather flow depths that vary through the wetland between 0.1 and 1.2 m (0.5 to 4 ft), depending on the types of vegetation planted, with the outlet structure designed so that the wetland can be periodically drawn down completely to dry the sediments; this provides for natural oxidation of built-up organics.

- Size the wet-weather storage volume using the methodology for ED ponds (discussed in sec 3.3.1.3) but with a maximum surcharge depth above the dry-weather flow depth of 0.6 m (2 ft) and a drawdown time of 24 h; this will reduce stress on herbaceous wetland plants. The 0.6 m depth limitation will determine the surface area required for the wetland.

- Design inlet structures to achieve sheet flow across the wetland to the maximum extent possible.

- Design the outlet structure to control the water surface and protect it from plugging by floatables common in wetlands.

- If open water is to be included in the wetland, it should be less than 50% of the total wetland area; the depth of the open water should follow the rules for the maximum permanent pool depth in retention ponds (WEF and ASCE, 1998).

In general, wetland designs are unique for each site and application. However, there are a number of geometric ratios and limiting depths for the design of a stormwater wetland that are recommended for adequate pollutant removal, ease of maintenance, and improved safety. Table 3-5 provides the recommended physical specifications and geometry for the various stormwater wetland design variants.

Table 3-5 Recommended Design Criteria for Stormwater Wetlands

Design Criteria	**Shallow Wetland**	**ED Shallow Wetland**	**Pond/ Wetland**	**Pocket Wetland**
Length-to-width ratio (minimum)	2:1	2:1	2:1	2:1

Extended-detention (ED)	No	Yes	Optional	Optional
Allocation of WQ_v volume (pool/marsh/ED) in %	25/75/0	25/25/50	70/30/0 (includes pond volume)	25/75/0
Allocation of surface area (deep water/low marsh/high marsh/semi-wet) in %	20/35/40/5	10/35/45/10	45/25/25/5 (includes pond surface area)	10/45/40/5
Forebay	Recommended	Recommended	Recommended	Optional
Micropool	Recommended	Recommended	Recommended	Recommended
Outlet configuration	Reverse-slope pipe or hooded broad-crested weir	Reverse-slope pipe or hooded broad-crested weir	Reverse-slope pipe or hooded broad-crested weir	Hooded broad-crested weir

Depth:
Deepwater: 1.5 to 6 ft below normal pool elevation
Low Marsh: 6 to 18 in. below normal pool elevation
High Marsh: 6 in. or less below normal pool elevation
Semi-wet zone: Above normal pool elevation
(Schueler, 1992a; Atlanta Regional Commission, 2001)

Depth

The stormwater wetland should be designed with the recommended proportion of the "depth zones." Each of the four wetland design variants has depth zone allocations which are given as a percentage of the stormwater wetland surface area. The four basic depth zones are given below.

Deepwater zone
The deep water zone is between 1.5 and 6 ft deep. It includes the outlet micropool and deepwater channels through the wetland facility. This zone supports little emergent wetland vegetation, but may support submerged or floating vegetation.

Low Marsh Zone
Low marsh zone is from 6 to 18 in. below the normal permanent pool or water surface elevation. This zone is suitable for the growth of several emergent wetland plant species.

High Marsh Zone
This is the zone from 6 in. below the pool to the normal pool elevation. This zone supports a greater density and diversity of wetland species than the low marsh zone and has a higher surface area to volume ratio.

Semi-wet Zone
Semi-wet zone refers to those areas above the permanent pool that are inundated during larger storm events and support a number of species that can survive flooding.

A minimum dry-weather flow path of 2:1 (length-to-width) is required from inflow to outlet across the stormwater wetland and should ideally be greater than 3:1. This path may be achieved by constructing internal dikes or berms, using marsh plantings, and by using multiple cells. Finger dikes are commonly used in surface flow systems to create serpentine configurations and prevent short-circuiting. Micro topography, or contours along the bottom of a wetland or marsh, which provides a variety of conditions for different species needs and increases the surface area to volume ratio, is encouraged to enhance wetland diversity.

A 4 to 6 ft deep micropool should be included in the design at the outlet to prevent the outlet from clogging and resuspension of sediments, and to mitigate thermal effects. In general, the maximum depth of any permanent pool areas should not exceed 6 ft.

The volume of the extended detention should not comprise more than 50% of the total WQ_v, and its maximum water surface water elevation should not extend more than 3 ft above the normal pool. Qp and/or Cpv storage can be provided above the maximum WQ_v elevation within the wetland.

The perimeter of all deep pool areas (4 ft or greater in depth) should be surrounded by safety and aquatic benches similar to those for stormwater ponds

The contours of the wetland should be irregular to provide a more natural landscaping effect.

Pretreatment/Inlets

A wetland facility should have a sediment forebay or upstream pretreatment in order to remove incoming sediment from the stormwater flow prior to dispersal into the wetland. The forebay should consist of a separate cell, formed by an acceptable barrier and should be provided at each inlet, unless the inlet provides less than 10% of the total design storm inflow to the wetland facility.

The forebay should be sized to contain 0.1 in. per impervious acre of contributing drainage and should be 4 to 6 ft deep. The pretreatment storage volume is part of the total WQv requirement and may be subtracted from WQv for wetland storage sizing.

A fixed vertical sediment depth marker should be installed in the forebay to measure sediment deposition over time and the bottom of the forebay may be hardened to make sediment removal easier.

Inflow channels need to be stabilized with flared riprap aprons, or the equivalent and inlet pipes to the pond can be partially submerged. Exit velocities from the forebay must be nonerosive.

Outlet Structures

Flow control from a stormwater wetland is typically accomplished with the use of a concrete or corrugated metal riser and barrel. While the riser is a vertical pipe or inlet structure that is attached to the base of the micropool with a watertight connection, the outlet barrel is a horizontal pipe attached to the riser that conveys flow under the embankment. The riser is recommended to be located within the embankment for maintenance access, safety and aesthetics.

A number of outlets at varying depths in the riser provide internal flow control for routing of the water quality, channel protection, and overbank flood protection runoff volumes. The number of orifices varies and is usually a function of the pond design. Alternative hydraulic control methods to an orifice can be used and include the use of a broad-crested rectangular, V-notch, proportional weir, or an outlet pipe protected by a hood that extends at least 12 in. below the normal pool.

Higher flows pass through openings or slots protected by trash racks further up on the riser or in a separate outlet. After entering the riser, flow is conveyed through the barrel and is discharged downstream. Anti-seep collars should be installed on the outlet barrel to reduce the potential for pipe failure. Riprap, plunge pools or pads, or other energy dissipators, are to be placed at the outlet of the barrel to prevent scouring and erosion. If a wetland facility daylights to a channel with dry weather flow, care should be taken to minimize tree clearing along the downstream channel, and to re-establish a forested riparian zone in the shortest possible distance.

The wetland facility should have a bottom drain pipe located in the micropool with an adjustable valve that can completely or partially de-water the wetland within 24 h. However, this requirement may be waived for coastal areas, where positive drainage is difficult to achieve due to very low relief.

The wetland drain should be sized one pipe size greater than the calculated design diameter. The drain valve is typically a handwheel activated knife or gate valve; valve controls shall be located inside of the riser at a point where they would not normally be inundated, and can be operated in a safe manner.

Emergency Spillway

An emergency spillway may be included in the stormwater wetland design to safely pass flows that exceed the design storm flows. It should be located so that downstream structures will not be affected by spillway discharges. One recommendation is to provide a minimum of 1 ft of freeboard, measured from the top of the water surface elevation for the extreme flood to the lowest point of the dam embankment, not counting the emergency spillway.

Maintenance Access

A maintenance right-of-way should be provided to the wetland facility from a public or private road. Maintenance access should be at least 12 ft wide, have a maximum slope of no more than 15%, and be appropriately stabilized to withstand maintenance equipment and vehicles. The maintenance access should extend to the forebay, safety bench, riser and outlet, and, to the extent feasible, be designed to allow vehicles to turn around. Access to the riser should be provided by lockable manhole covers and manhole steps within easy reach of valves and other controls.

Vegetation

The vegetation diversity in a constructed wetland is established by the landscape plan or volunteer vegetation. The selection of vegetation should be limited to native plant species suitable for the pool depths expected within the different depth zones (VA DCR, 1999). Care should be taken to avoid introducing exotic or invasive species. This problem can be overcome by the use of appropriate donor soil and wetland mulch. Suitable plants for created wetlands vary between different eco-regions. However, the wetland plants chosen for created wetlands should incorporate the following attributes:

- tolerance to wide ranges of water elevations, salinity, temperature, and pH,
- a mixture of perennials and annuals,
- moderate amounts of leaf production, and
- proven removal efficiencies, e.g., *Scriptus* spp. (WEF and ASCE, 1998).

Additional considerations include:

- The use of vegetation and an appropriate landscaping plan should provide elements that promote greater wildlife and waterfowl use within the wetland and buffers.

- Woody vegetation may not be planted on the embankment or allowed to grow within 15 ft of the toe of the embankment and 25 ft from the principal spillway structure.

- A wetland buffer shall extend 25 ft outward from the maximum water surface elevation, with an additional 15 ft setback to structures. The wetland buffer should be contiguous with other buffer areas that are required by existing regulations, or are part of the overall stormwater management concept plan. No structures shall be located within the buffer and an additional setback to permanent structures may be provided.

- Existing trees should be preserved in the buffer area during construction. It is desirable to locate forest conservation areas adjacent to ponds. Resident goose populations can be discouraged by planting the buffer with trees, shrubs and native ground covers.

Figure 3-3 shows a schematic of a stormwater wetland (CASQA, 2003).

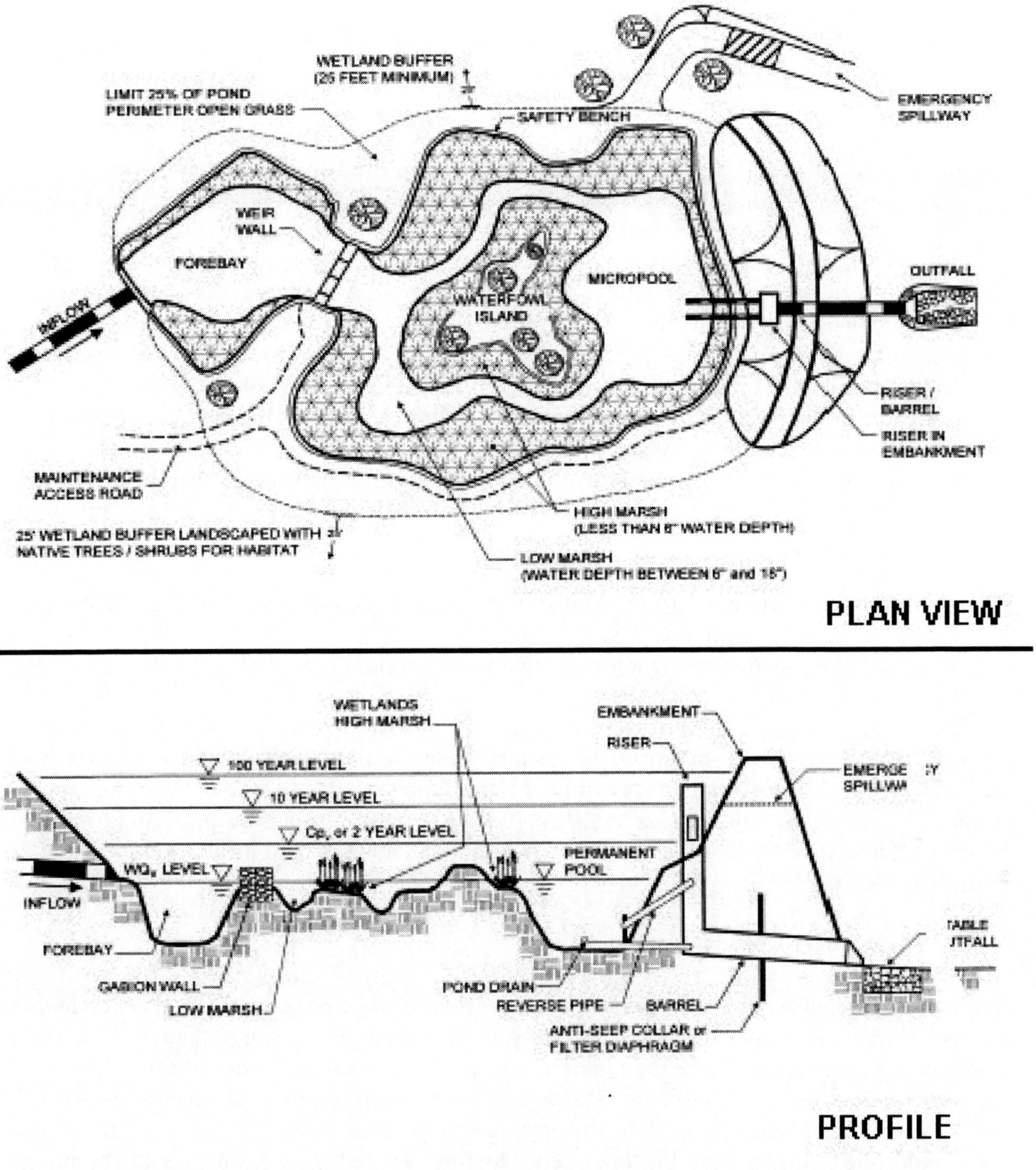

Figure 3-3. **Schematic of a Constructed Wetland** (CASQA, 2003)

3.3.4 Grassed Swales

The term "grassed swales", also known as grassed water courses or vegetated swales, refers to the use of grassed conveyances, which are essentially earthen channels vegetated with erosion resistant and flood-tolerant grasses. They are designed to infiltrate runoff from intermittent storm events or to transfer rainfall excess at a non-erosive velocity to desired locations for retention, detention, storage, or discharge. There are three variations of grassed swales: (1) traditional grass swales; (2) grass swales with a media filter; and, (3) wet swales. Although using grassed swales for the sole purpose of conveying stormwater has become a common practice in residential and institutional settings, their effective use for water quality control is a fairly recent practice, and in most cases is better accomplished in combination with other BMPs placed downstream to meet stormwater management requirements (Yousef *et al.*, 1985; MDE, 2000; CASQA, 2003). With respect to total stormwater management, the desirable attributes in vegetated grass swales include:

- slower flow velocities than pipe systems, which result in longer times of concentration and corresponding reduction of peak discharges,
- ability to disconnect directly connected impervious surfaces, such as driveways and roadways, thus reducing the computed runoff curve number (CN) and reduction of peak discharge,
- filtering of pollutants by grass media,
- infiltration of runoff into the soil profile, thus reducing peak discharges and providing additional pollutant removal, and
- uptake of pollutants by plant roots (phytoremediation) (WEF and ASCE, 1998).

A water quality swale is appropriate where greater pollutant removal efficiency is desired; the capacity to accept runoff from large design storms being limited in swales, these treatment swales must often lead into storm drain inlets to prevent large, concentrated flows from gullying/eroding the swale. Placing check dams across the flow path to temporarily pond runoff could improve the hydrologic performance of swales with regard to flow attenuation and infiltration for small design storms (Schueler, 1987).

3.3.4.1 Stormwater Control

Grassed swales and water quality swales usually provide some peak attenuation depending on the storage volume created by the check dams. However, flood control should be considered a secondary function of grassed swales since the required storage volume for flood control is usually more than the swales can provide (NVPDC and ESI, 1992). Swales act to control peak discharges in two ways:

- Reduction in runoff velocity by the grass, which depends on the length and slope of the swale, which in turn, lengthens the time needed for runoff to reach the desired control point, and can at least partially attenuate the post development peak discharge rate.
- A portion of the runoff passing through the swale infiltrates into the soil and does not appear at the downstream control point. However, this seldom exceeds a few tenths of an inch and depends on soils and slope besides the short contact time of runoff with the swale (5-20 min.); swale soils have less infiltration capacity than undisturbed soils as they are heavily compacted to achieve the desired slope and load bearing capacity. Due to previous saturation of the

swale soils by the same rain that supplies runoff to the swale, infiltration rates in a swale will almost always be near the minimum rates for the local soil type.

3.3.4.2 Pollutant Removal Capability

The primary pollutant removal mechanisms associated with grassed swales are sedimentation and infiltration into the subsoil. Adsorption and filtration mechanisms can be considered as secondary removal mechanisms (Schueler, 1987). Changes in the flow hydraulics affected by routing the flow through grassed channels increase the opportunity for infiltration of soluble pollutants, deposition of suspended solids, filtration of suspended solids by vegetation, and adsorption of soluble particles by plants. The flow rate becomes a critical design element since surface runoff must pass slowly through the filter to provide sufficient contact time for the aforementioned removal mechanisms to function effectively (NVPDC and ESI, 1992; WY DEQ, 1999).
Conventional grassed swale designs have achieved mixed performance in removing particulate pollutants such as suspended solids and trace metals. They are generally unable to remove significant amounts of soluble nutrients. Design practices that increase the retention time of urban runoff will increase removal efficiencies for soluble forms of nitrogen and phosphorus (Yousef *et al.*, 1985). Biofilters that increase detention, infiltration, and wetland uptake within the swale have the potential to substantially improve swale removal rates (CASQA, 2003).

Positive factors influencing pollutant removal (WY DEQ, 1999; CASQA, 2003) include:

- check dams,
- low slopes,
- permeable subsoils and soil moisture holding capacity,
- dense grass cover or vegetation or mulches,
- long contact time,
- smaller storm events,
- coupling swales with plunge pools,
- infiltration trenches or pocket wetlands, and
- swale length greater than one hundred ft.

Negative factors influencing pollutant removal include:

- compacted subsoils,
- short runoff contact storms,
- large storm events,
- snow melt events,
- short grass heights,
- steep slope (6% or greater),
- runoff velocities greater than 1.5 fps,
- peak discharge greater than 5 cfs, and
- dry-weather flow.

3.3.4.3 Design Considerations

The basic design procedure for a swale system was developed by Chow (Chow, 1959) and has been used in a number of ways in sizing and designing grass swales of varying degrees of complexity and design robustness. A summary of the various approaches to the design procedure is presented in this book. Reported estimates of low pollutant removal efficiencies for grassed swales verify the need to improve standard design procedures to make them more effective for BMP purposes. Studies on swale performance are ambiguous making it hard to propose specific estimates for swale pollutant removal efficiency. The design of a grassed swale includes calculations for traditional swale parameters such as flow rate, maximum permissible velocities, etc. along with storage volume calculations for the water quality volume (VA DCR, 1999). A moderate removal of particulate pollutants can be achieved during small storms if a swale conforms to the design considerations discussed below.

General Design Recommendations

A dry swale system consists of an open conveyance channel with a filter bed of permeable soils that overlays an underdrain system. Flow passes into and is detained in the main portion of the channel where it is filtered through the soil bed. Runoff is collected and conveyed by a perforated pipe and gravel underdrain system to the outlet. A wet swale consists of an open conveyance channel that has been excavated to the water table or to poorly drained soils. Check dams are used to create multiple wetland "cells" that act as miniature shallow marshes. The dry and the wet swale are designed to treat the WQ_v through a volume-based design, and to safely pass larger storm flows. Runoff enters the channel through a pretreatment forebay or along the sides of the channel as sheet flow through the use of a pea gravel flow spreader trench along the top of the bank.

Siting

The suitability of a swale at a site will depend on land use, size of the area served, soil type, slope, imperviousness of the contributing watershed, and dimensions and slope of the swale system (Schueler *et al.*, 1992). In general, swales (dry/wet) should be sited such that the topography allows for the design of a channel with sufficiently mild slope and cross-sectional area to maintain non-erosive velocities (Atlanta Regional Commission, 2001).

Swale siting should also take into account the location and use of other site features such as buffers, undisturbed natural areas, and natural drainage courses, and should attempt to aesthetically "fit" the facility into the landscape (CASQA, 2003).

Roadside ditches should be regarded as significant potential swale/buffer strip sites and should be utilized as such whenever possible. If flow is to be introduced through curb cuts, it is recommended to place pavement slightly above the elevation of the vegetated areas. Curb cuts should be at least 12 in. wide to prevent clogging (WEF and ASCE, 1998).

Dry swales can be sited on most soils; however, native soils with low permeability need to be amended or replaced to increase infiltration. A wet swale can be used where the water table is at or near the soil surface, or where there is a sufficient water balance in poorly drained soils to support a wetland plant community (Metro Council, 2001).

The soil below the swale should not consist of too much gravel or coarse sand, as these constituents do not easily support dense vegetation. It should be undisturbed, as this area may be periodically inundated and remain wet for long periods of time. In areas with steep slopes, swales should be employed in locations where they can be parallel to the contours. Unless existing soils are highly permeable, they should be replaced with a 30 in. depth of a sand/soil mixture (approximately 50/50 mix) to ensure infiltration.

An underlying engineered soil bed and underdrain system may be utilized in areas where the soils are not permeable and the swale would remain full of water for extended periods of time (creating nuisance conditions). This soil bed should consist of a moderately permeable soil material with a high level of organic matter; e.g., 50% sand, 20% leaf mulch, 30% top soil. The soil bed should be 30 in. deep and accompanied by a perforated pipe and gravel underdrain system. In residential developments with marginal soils, it may be appropriate to provide a soil bed and underdrain system in all grassed swales to avoid possible safety and nuisance concerns.

Appropriate soil stabilization methods, such as mulch, blankets, or mats should be used before establishing vegetation. Seeding, sodding, and other items related to establishing vegetation should be in accordance with accepted erosion control and planting practices (WEF and ASCE, 1998; Metro Council, 2001).

Physical Specifications/Geometry

The detention/retention capacity of grassed swales is governed by the runoff associated with the "water quality storm." The swale length, width, depth, and slope should be designed to temporarily accommodate the WQ_v through surface ponding. The WQ_v is retained for 24 h, but ponding may continue indefinitely depending on the depth and elevation to the watertable. The WQ_v for high density residential, commercial, and industrial land uses will most likely be too high to be accommodated with most swale designs, and swales in these cases may be appropriate for pretreatment in association with other practices for these higher density land uses (Metro Council, 2001). The swale can be sized as both a treatment facility for the design storm and as a conveyance system to pass the peak hydraulic flows of the 100-yr storm if it is located online (CASQA, 2003).

Water Quality Volume

The purpose of a grassed swale used as a conveyance channel is to transport stormwater to the discharge point. However, the purpose of a water quality grassed swale is to slow the water as much as possible to encourage pollutant removal. The use of check dams will create segments of the swale that will be inundated for a period of time. The required total storage volume behind the check dams is equal to the water quality volume for the contributing drainage area to that point. However, the maximum ponding depth behind the check dams should not exceed 18 in. To insure that this practice does not create nuisance conditions, an analysis of the subsoil is recommended to verify its permeability.

Swale Geometry

A grassed swale should have a trapezoidal cross-section to spread flows across its flat bottom. Triangular or parabolic shaped sections are generally not recommended as they tend to concentrate the runoff. However, a parabolic shape could be acceptable provided the width is equal to or greater than the design bottom width for a trapezoidal cross section (Atlanta Regional Commission, 2001; Metro Council, 2001). The side slopes of the swale should be no steeper than 3H:1V to simplify maintenance and help prevent erosion (Schueler, 1987; WA DOE, 2001).

Bottom Width

The bottom width of the swale should be 2 ft minimum and 8 ft maximum in order to maintain sheet flow across the bottom and avoid concentration of low flows. The 2 ft requirement would allow for construction considerations and would ensure a minimum filtering surface for water quality treatment, while the 8 ft maximum would reduce the likelihood of flow channelization within a portion of the bottom of the swale. Widths up to 16 ft may be used if separated by a dividing berm or structure to avoid braiding (WEF and ASCE, 1998; VA DCR, 1999; Metro Council, 2001). The actual design width of the swale is determined by the maximum desirable flow depth, as discussed below.

Flow Depth

The flow depth for a water quality grassed swale should be approximately the same as the height of the grass. An average grass height for most conditions is 4 in. Therefore, the maximum flow depth for the water quality volume should be 4 in. (CWP, 1996b). According to (WEF and ASCE, 1998), the maximum depth of flow should not be greater than one-third of the gross or emergent wetland vegetation height for infrequently mowed swales or not greater than one-half of the vegetation height for regularly mowed swales, up to a maximum of approximately 75 mm (3 in.) for grass and approximately 50 mm (2 in.) below the normal height of the shortest wetland plant species in the biofilter.

Flow Velocity

The maximum velocity of the water quality volume through the grassed swale should be no greater than 1.5 fps. The maximum design velocity of the larger storms should be kept low enough so as to avoid resuspension of deposited sediments. The 2-yr storm recommended maximum design velocity is 4 fps and the 10-yr storm recommended maximum design velocity is 7 fps.

Longitudinal Slope

The slope of the grassed swale should be as flat as possible, while maintaining positive drainage and uniform flow, to permit the temporary ponding of the WQ_v within the channel without having excessively deep water at the downstream end. The minimum constructable slope is between 0.75 and 1.0% and the maximum slope depends upon what is needed to maintain the desired flow velocities as well as to provide adequate storage for the water quality volume while avoiding excessively deep water at the downstream end. Generally, a slope of between 1 and 3% is recommended. (CASQA, 2003) recommends a maximum of 2.5%, (WEF and ASCE, 1998; Metro Council, 2001) recommend 2% and (VA DCR, 1999) recommends 3%. The slope should never exceed 5%.

Swale Length

Swale length is dependent on the swale geometry and the ability to provide the required storage for the water quality volume. However, the swale should have a length that provides a minimum hydraulic residence time of at least 10 minutes (CASQA, 2003)), and regardless of the recommended detention time, the swale should be not less than 100 ft in length.

Swale Capacity

The capacity of the grassed swale is a combined function of the flow volume (the water quality volume) and the physical properties of the swale such as longitudinal slope and bottom width. The depth of flow and velocity for any given set of values can be obtained by using the Manning equation or channel flow nomographs. The Manning's 'n' value, or roughness coefficient, varies with the depth of flow and vegetative cover. An n value of 0.15 is considered appropriate for flow depths of up to 4 in.(equal to the grass height). The n value decreases to a minimum of 0.03 for grass swales at a depth of approximately 12 in. A grassed swale should have the capacity to convey the peak flows from the 10-yr design storm without exceeding the maximum permissible velocities. (It must be noted that a maximum velocity is specified for the 2-yr and 10-yr design storms to avoid resuspension of deposited sediments and other pollutants and to prevent scour of the channel bottom and side slopes). The swale should pass the 10-yr flow over the top of the check dams with 6 in. minimum of freeboard. Alternatively, a bypass structure may be engineered to divert flows from the larger storm events (runoff greater than the water quality volume) around the grassed swale. However, when the additional area and associated costs for a bypass structure and conveyance system are considered, it may be more economical to simply increase the bottom width of the grassed swale. It should then be designed to carry runoff from the 10-yr design storm at the required permissible velocity. The Manning equation can be used to adjust the longitudinal slope and bottom width to achieve the maximum allowable velocity (VA DCR, 1999). The following criteria are probably most applicable in warm and temperate non-semi-arid climates and should be met or exceeded during the biofiltration capacity design event (WEF and ASCE, 1998): "maximized" runoff hydraulic residence time of 5 min or more; maximum flow velocity less than 0.3 m/s (0.9 ft/sec); Manning's n = 0.20 for routinely mowed swales, and Manning's n = 0.24 for infrequently mowed swales.

One recommended procedure for designing grassed swales (Claytor and Schueler, 1996) outlines a series of guidelines as shown below and in Table 3-6:

- Compute the water quality treatment volume (WQ_v) for the given land surfaces as required by the local permitting agency.
- Identify the required swale bottom width, depth, length and slope necessary to store the WQ_v within a shallow ponding depth (18 in. maximum).
- Compute the WQ_v drawdown time to ensure that it is less than 24 h.
- Compute the 2-yr and 10-yr frequency storm event peak discharges.
- Check the 2-yr velocity for erosive potential (adjust swale geometry, if necessary, and reevaluate WQ_v design parameters).
- Check the 10-yr depth and velocity for capacity (adjust swale geometry, if necessary, and reevaluate WQ_v design parameters).
- Provide minimum freeboard above 10-yr stormwater surface profile (6-in. minimum recommended).

Table 3-6. Design Criteria for Dry (and Wet) Swale Systems

Parameter	**Swale Design Criteria**
Pretreatment volume	.05 in. per impervious acre, at initial flow point
Preferred shape	Trapezoidal or parabolic
Bottom width	2 ft minimum, 8 ft maximum; widths up to 16 ft are allowable if a dividing berm or structure is used
Side slopes	2:1 maximum, 3:1, or flatter preferred
Longitudinal slope	1.0% to 2.0% without check dams
Sizing criteria	Length, width, depth and slope needed to provide surface storage for WQ_v. Outlet structures, when used, should be sized to release WQ_v over 24 h
Underlying soil bed	Equal to swale width Dry Swale: Moderately permeable soils, 30 in. deep with gravel/pipe underdrain system if needed Wet Swale: Undisturbed soils, no underdrain system
Depth and capacity	Surface storage of WQ_v with a maximum depth of 18 in. for water quality treatment (12in. average depth) Safely convey 2 yr storm with non-erosive velocity ($\leq$ 4.0 ft/s) Adequate capacity for 10 yr storm with 6 in. of freeboard

(Claytor and Schueler, 1996)

Pretreatment/Inlets

Inlets to swales must be provided with erosion controls as needed (e.g., riprap, flow spreaders, energy dissipators, sediment forebays, etc.).

Pretreatment of runoff in both dry and wet swale system is typically provided by a sediment forebay located at the inlet. The pretreatment volume should be equal to 0.1 in. per impervious acre. A forebay large enough to accommodate 25% of the water quality volume is created by installing a check dam, constructed of timber or concrete, between the inlet and the main body of the swale and/or driveway crossings (Atlanta Regional Commission, 2001; Metro Council, 2001). The checkdam should overlay a stone base to prevent downstream scour. The area downstream of the checkdam should be protected from scouring with riprap or channel lining. In the undesirable event of clogging in the surface soils, a checkdam may also be installed at the downstream end of the swale, along with an optional pea gravel window to route water to the underdrain.

Enhanced swale systems that receive direct concentrated runoff may have a 6 in. drop to a pea gravel diaphragm flow spreader at the upstream end of the control. A pea gravel diaphragm and gentle side slopes should be provided along the top of the channels to provide pretreatment for lateral sheet flows.

Check dams

Check dams are utilized in swales for two reasons: to increase pollutant removal efficiency and/or to compensate for steep longitudinal slope. The dams should be installed perpendicular to the direction of flow and anchored into the slope of the channel. The side slopes of the check dams should be between 5 and 10 to 1 to facilitate mowing operations. The berm height should not exceed 0.6 m (2 ft), and water ponded behind the berm should infiltrate into the soils within 24 h (UDFCD, 2002). Check dams should be spaced so that the toe of the upstream dam is at the same elevation as the top of the downstream dam. For best performance, check dams should have a level upper surface rather than the uneven surface of a riprap check dam. Earthen check dams are also not recommended due to erosion potential and high maintenance effort.

Level Spreaders

Level spreaders are diminutive check dams used to provide a uniform flow distribution across the swale bottom. The hydraulic design of the swale assumes a uniform distribution, which is difficult to attain without the aid of level spreading devices. The device, placed at the swale inlet, may consist of a shallow weir across the channel bottom, a stilling basin, or perforated pipe. A sediment clean-up area should be provided for ease of maintenance.

Flow Bypass

Flow bypass should be considered for high flow events to avoid erosion and channelization. Flow bypass also allows diversion of flows during swale maintenance, regrading, and vegetation establishment. Flow can be bypassed by installing a pipe parallel to the swale and a flow regulating device inside the inlet structure. High flow bypasses may be of two types: "first-flush" treatment or design flow treatment. The "first-flush" treatment is based on the principle that storm event pollutants are more concentrated during the "first-flush." Biofiltration swales can be designed for treating stormwater only from this initial portion of the storm event, and would require bypassing stormwater flow around the swale during higher portions of flow. More typically, swale bypasses are designed to treat the design flow throughout the storm event, bypassing only the flows in excess of the design flow.

Riprap

Riprap is used as an energy dissipation or erosion control device in grassy swales. Riprap pads, consisting of 152 to 228 mm (6 to 9 in.) rocks that fit tightly across the bed may be used as an energy dissipater at the swale inlet and continuing for a distance of 1.5 to 3 m (5 to 10 ft) downstream. Riprap can also be used to line the swale channel if erosion and/or channelization of the swale bottom are of concern. Riprap could also be used with check dams as described above.

Outlet Structures

Discharges from grassed swales must be conveyed at non-erosive velocities to either a stream or a stabilized channel to prevent scour at the outlet of the swale. In dry swales, the underdrain system should discharge to the storm drainage infrastructure of a stable outfall; in wet swales, outlet protection must be used at any discharge point to prevent scour and downstream erosion.

Emergency Spillway

Enhanced swales must be adequately designed to safely pass flows that exceed the design storm flows.

Landscaping and Vegetation

Landscape design should specify proper grass species and wetland plants based on specific site, soil, and hydric conditions present along the channel. A dense cover of water-tolerant, erosion-resistant grass or other vegetation must be established. Grasses used in swales should have the following characteristics:

- a deep root system to resist scouring,
- a high stem density, with well-branched top growth,
- tolerance to flooding,
- resistance to being flattened by runoff, and
- an ability to recover growth following inundation.

Recommended grasses include but are not limited to the following: Kentucky-31, tall fescue, reed canary grass, redtop, rough-stalked blue grass, switch grass, little blue stem, and big blue stem. It should be noted that these grasses can be mixed.

The selection of an appropriate vegetative lining for a grassed swale is based on several factors including climate, soils, and topography (VA DCR, 1999). Erosion control matting should be used to stabilize the soil before seed germination. This protects the swale from erosion during the germination process. In most cases, the use of sod may be required to provide immediate stabilization on the swale bottom and/or side slopes (WEF and ASCE, 1998). Figures 3-4 and 3-5 provide a representative typical section including both a cross-section and plan view of a dry and wet swale respectively (MDE, 2000).

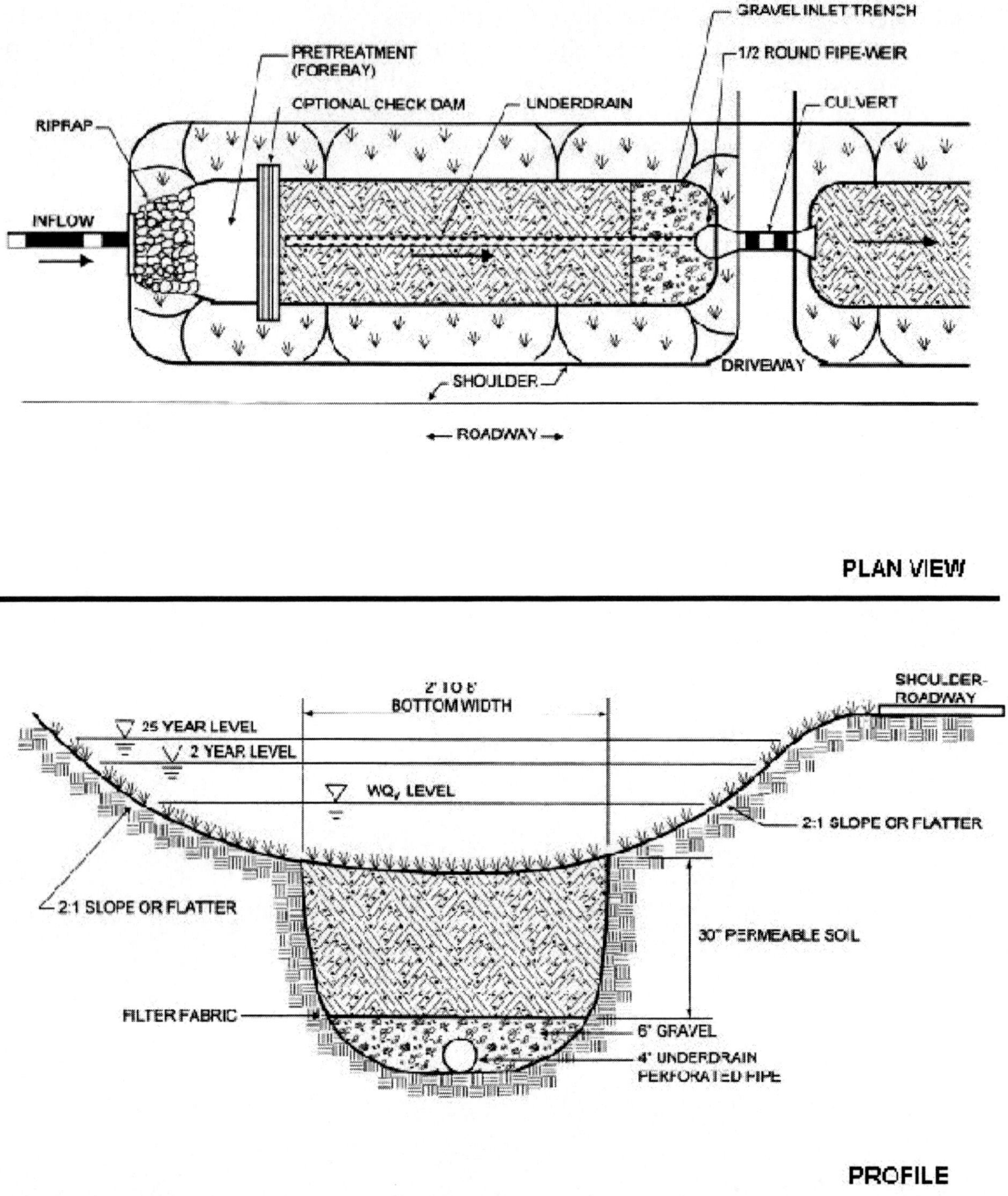

Figure 3-4 Schematic of a Dry Swale (MDE, 2000)

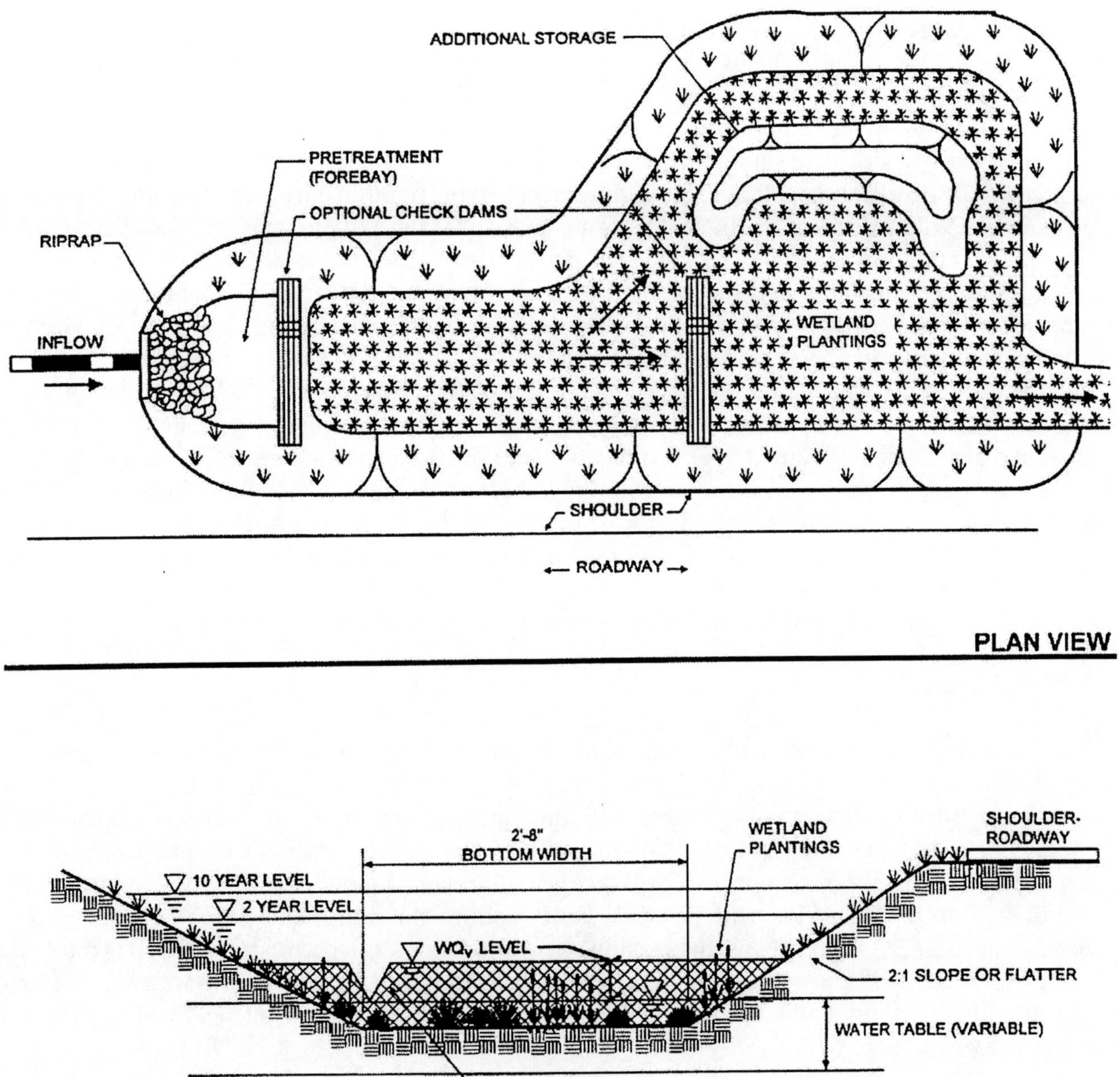

Figure 3-5 Schematic of a Wet Swale (MDE, 2000)

3.3.5 Vegetated Filter Strips (VFS)

Filter strips are uniformly graded and densely vegetated sections of land, engineered and designed to treat runoff and remove pollutants through vegetative and soil filtering, evapotranspiration, and infiltration. Filter strips are similar to grassed swales in many respects, except that they are designed to accept only overland sheet flow. The requirement that the runoff from an adjacent impervious area be evenly distributed across the filter strip is not an easy task due to the strong tendency of runoff to concentrate and form a channel, reducing the performance efficiency of the filter strip and in some cases leading to erosion of portions of the filter strip (Schueler, 1987). VFS are best suited to treat runoff from roads and highways, roof down spouts, very small parking lots, and pervious surfaces. They can function as the outer zone of a stream buffer, or as a pretreatment for other structural stormwater controls (WY DEQ, 1999; Atlanta Regional Commission, 2001). To function properly, a filter strip must be equipped with some level spreading device, densely vegetated with a mix of erosion resistant plant species that effectively bind the soil, graded to a uniform, even, and relatively low slope, and at least as long as the contributing runoff area (Knoxville, 2003). There are two filter strip designs: a simple filter strip and another design that includes a permeable berm at the bottom. The presence of the berm increases the contact time with the runoff, thus reducing the overall width of the filter strip required to treat stormwater runoff. As filter strips are typically an on-line practice, they must be designed to withstand the full range of storm events without eroding.

3.3.5.1 Stormwater Control

Filter strips do not provide enough storage or infiltration to effectively reduce peak discharges to pre-development levels for design storms (Schueler, 1987; NVPDC and ESI, 1992). The lowering of runoff velocities and runoff volume, observed sometimes in VFS, may not be typically adequate for controlling stream channel erosion or flooding (NVPDC and ESI, 1992). Little attenuation of peak runoff rates and volumes is observed for larger events, depending on soil properties, suggesting the practice of following strips with another BMP option that can reduce flooding and erosion downstream (CASQA, 2003). The increasing use of filter strips as a pretreatment BMP in integrated stormwater management systems helps lower runoff velocities and hence the watershed time of concentration, slightly reduce both runoff volumes and watershed imperviousness, and contribute to groundwater recharge.

3.3.5.2 Pollutant Removal Capability

Both swales and filter strips exhibit similar mechanisms of pollutant removal. Pollutant removal from filter strips is highly variable and depends primarily on density of vegetation and contact time for filtration and infiltration and soil moisture absorption capacity (Atlanta Regional Commission, 2001). The mechanisms include the filtering action of vegetation, deposition in low velocity areas, or by infiltration into the subsoil. The rate of removal appears to be a function of the length, slope and soil permeability of the strip, the size of the contributing runoff area, and the runoff velocity (Schueler, 1987).

Vegetated buffer strips are generally effective in reducing the volume and mass of pollutants in runoff and when designed properly, tend to provide somewhat better treatment of stormwater runoff than swales with fewer tendencies for flow concentration and the resulting erosion (CASQA, 2003). Filter strips can effectively reduce particulate pollutant levels in areas where runoff velocity is low to moderate; however, the ability to remove soluble pollutants under the same conditions is highly variable (WY DEQ, 1999). Soluble pollutants in filter strips are removed by pollutant infiltration into the soil and subsequent uptake by rooted vegetation. However, the efficiency of soluble pollutant removal may not be high since only a small portion of the incoming runoff will be infiltrated (Schueler, 1987). Filter strips are effective in removing particulate pollutants such as sediment, organic matter, and trace metals as observed from results from small test plots and several modeling studies (Schueler, 1987). They also exhibit good removal of litter and other floatables as the water depth in these systems is well below the vegetation height. Forested filter strips appear to have greater pollutant removal capability than grass filter strips because of greater uptake and long-term retention of nutrients in forest biomass.

Positive factors influencing pollutant removal include:

- minimum strip width of fifty ft,
- slope of 5% or less,
- clay soil or organic matter surface,
- contributing area of less than 5 acre,
- grass height of 6 to 12 in., and
- sheet flow.

Negative factors influencing pollutant removal include:

- runoff velocity > 2.5 fps, depending on site conditions (Horner, 1988),
- slopes greater than 15%,
- hilly terrain, and
- unmowed filter strips.

3.3.5.3 Design Considerations

General Considerations

VFS have limited feasibility as a water quality control in ultra-urban settings with a high percentage of impervious area where runoff velocities and peak discharge rates are high and flow is concentrated. Their use is therefore primarily restricted to low and medium density residential areas (16 to 21% impervious) where they can accept rooftop runoff and runoff from pervious areas such as lawns, or as a pre-treatment component for structural BMPs in higher density developments (WY DEQ, 1999). The retrofit capability is relatively simple if enough land area is available to adequately service the contributing watershed area, and soil and slope conditions are favorable.

Filter strips should be constructed outside the natural stream buffer area whenever possible to maintain a more natural buffer along the streambank (Atlanta Regional Commission, 2001).

Forests and other natural areas should not be destroyed to create a filter strip system, as such areas may already be functional or may only need to be enhanced to function properly as treatment systems. Disturbance of native vegetation in buffer areas should be avoided whenever possible (Metro Council, 2001). Other considerations include:

- Adequate pollutant removal may not be observed on slopes over 15%; filter strips require climates that can sustain vegetative cover on a year-round basis; contributing upland area must be small (1 to 5 acres) so that runoff arrives at the filter strip as overland sheet flow; use of native vegetation or vegetation appropriate for the local climate is essential to enhance plant survival (WY DEQ, 1999; CASQA, 2003).

- Filter strips should not be used on soils that cannot sustain a dense grass cover with high retardance. It is recommended to choose grasses that can withstand relatively high velocity flows at the entrances during both dry and wet periods (Atlanta Regional Commission, 2001).

Siting

The use of buffer strips is limited to gently sloping areas where the vegetative cover is robust and diffuse and where shallow flow characteristics are possible. Slopes should not exceed 15% or be less than 1% (CASQA, 2003). The vegetative surface should extend across the full width of the area being drained. The upstream boundary of the filter should be located contiguous to the developed area. The following site conditions should be considered when selecting a vegetated filter strip as a water quality BMP:

Vegetated filter strips should be used with soils having an infiltration rate of 0.52 in./h; (sandy loam, loamy sand). Soils should be capable of sustaining adequate stands of vegetation with minimal fertilization (VA DCR, 1999). The ability to remove nutrients from surface runoff improves where clay soils or organic matter are present (WY DEQ, 1999).

A shallow or seasonally high groundwater table will potentially inhibit infiltration and one recommendation from the VA DCR (VA DCR, 1999) is to have the lowest elevation in the filter strip at least 2 ft above the water table. Filter strips should be separated from the ground water by between 2 and 4 ft to prevent contamination and should not remain wet between storms (CASQA, 2003). Greater removal of soluble pollutants can be achieved where the water table is within 3 ft of the surface, i.e., within the root zone (WY DEQ, 1999). If the soil permeability and/or depth to water table are unsuitable for infiltration, the primary function of the filter strip becomes one of filtering and settling of pollutants. This requires a modified design to allow ponding of the water quality volume at the downstream end of the filter (VA DCR, 1999). Ponding area may be created by constructing a small permeable berm using a select soil mixture, with the maximum ponding depth behind the berm as 1 ft.

A natural area that is designed to serve as a vegetated filter strip should not be used for temporary sediment control.

Flows in excess of design flow must be ensured to move across or around the strip without damaging it. Higher flows can be handled by a bypass channel or overflow spillway with protected channel section (Atlanta Regional Commission, 2001).

Physical Specifications/Geometry

Slope

Filter strips appear to be a minimal design practice because they are basically no more than a grassed slope. The general requirement is that the slope should not exceed 15% and the slope should be at least 15 ft long to provide water quality treatment. Minnesota urban small sites BMP manual (Metro Council, 2001) recommends that filter strip slopes should be no less than 1 or 2% and no greater than 6%. Greater slopes will encourage concentrated flow and flatter slopes may result in ponding. Both the top and toe of the slope should be as flat as possible to encourage sheet flow and prevent erosion. The top of the strip should be installed 2 to 5 in. below the adjacent pavement, so that vegetation and sediment accumulation at the edge of the strip does not prevent runoff from entering. The flat cross-slope in filter strips ensures that runoff remains as sheet flow while filtering through the vegetation (CASQA, 2003).

Length

The filter strip should stretch the entire length of the impervious surface from where stormwater originates and when adjacent to a natural water body, it should stretch the entire length of the property of shoreline (Metro Council, 2001). VFS should be long enough to provide filtration and contact time for water quality treatment. Although smaller lengths for e.g., 15 ft, 25 ft are recommended in some states, (Atlanta Regional Commission, 2001), the recommended minimum length can be 50 to 75 ft based on the assumption that runoff changes from sheet flow to shallow concentrated flow after traveling 150 ft over pervious surfaces and 75 ft over impervious surfaces (CWP, 1996b); an additional 4 ft for any one percent increase in slope is recommended. Filter strips should be used to treat small drainage areas. Flow must enter the strip as sheet flow spread out over the width (longer dimension normal to the direction of flow) of the strip, generally no deeper than 1 to 2 in. However, the length is normally dictated by design method and the minimum length of a filter strip is:

$$W_{fMIN} = Q / q \qquad \textbf{(3-7)}$$

where
W_{fMIN} = minimum filter strip width perpendicular to flow (ft) (Atlanta Regional Commission, 2001).

The following section highlights the recommendations and minimum design guidelines for vegetated filter strips intended to enhance water quality. The responsibility rests with the designer to decide criteria applicable to each facility with any design modifications as well as to provide for the long-term functioning of the BMP.

Maximum discharge loading per ft of filter strip width (perpendicular to flow path) is found using Manning's equation:

$$q \frac{0.00236}{n} Y^{5/3} S^{1/2} \qquad \textbf{(3-8)}$$

where
q = discharge per ft of width of filter strip ($ft^3/s/ft$);
Y = allowable depth of flow (in.);
S = slope of filter strip (%); and
n = Manning's "n" roughness coefficient.
0.15 for medium grass, 0.25 for dense grass, 0.35 for very dense Bermuda-type grass. Compliance with the design parameters will result in optimal filter strip performance (NVPDC and ESI, 1992)

Width

Filter strips must be at least 15 ft wide in the direction of flow in order to be effective, however greater widths will enhance treatment. The steeper the slope, the wider the strip should be. NRCS recommends a minimum of 150 ft of filtering buffer between a land disturbance activity and a water body; depending upon soil types and slopes, it may be even greater (Metro Council, 2001).

The width of the filter strip should generally be equal to the width of the contributing drainage area and when this is not practical, a level spreader should be used to reduce the flow width to that of the filter strip. The width of the level spreader will determine the depth of flow and runoff velocity of the stormwater as it passes over the spreader lip and into the filter strip. While a wide lip will distribute the flow over a longer level section, it reduces the potential for concentrated flows across the filter (VA DCR, 1999).

A level spreader should be provided at the upper edge of a filter strip when the width of the contributing drainage area is greater than that of the filter and may extend across the width of the filter, leaving only 10 ft open on each end. Many configurations of the level spreader can be used and include a concrete sill or weir, curb stops, curb and gutter with "saw teeth" cut into it, or a level trench (12 in wide by 24 in deep), filled with pea gravel or crushed stone; the key is to have a long, continuous, and level overflow elevation to spread the concentrated flow into sheet flow upstream of the filter strip (VA DCR, 1999; Atlanta Regional Commission, 2001). An effective technique is to use a pea gravel diaphragm at the top of the slope (a small trench running along the top of the filter strip) which serves two purposes: (i) it acts as a pretreatment device settling out sediment particles before they reach the practice; and (ii) it acts as a level spreader, maintaining sheet flow as runoff flows over the filter strip.

Pervious Berm

A pervious berm may be installed to force ponding in a vegetated filter strip. A pervious berm of sand and gravel, or soils meeting USDA sandy loam or loamy sand texture, or any other moderately permeable soil could be installed at the toe of the slope to enhance the effectiveness of the filter strip. This could also include outlet pipes flowing through the berm or an overflow weir to provide an area for temporary shallow ponding and accommodate a portion or all of the water quality volume.

A pervious berm may be installed to force ponding in a vegetated filter strip. It should be constructed using a moderately permeable soil such as ASTM ML, SM or SC. Soils meeting USDA sandy loam or loamy sand texture, with a minimum of 10 to 25 % clay, may also be used. Additional loam should be used on the berm ± 25% to help support vegetation. An armored overflow should be provided to allow larger storms to pass without overtopping the berm. Maximum ponding depth behind a pervious berm is 1 ft (VA DCR, 1999; Metro Council, 2001).

Vegetation

A filter strip should be densely vegetated with a mix of erosion resistant plant species that effectively bind the soil. The selection of plants should be based on their compatibility with climate conditions, soils, and topography and their ability to tolerate stresses from pollutants, variable soil moisture conditions, and ponding fluctuations. A filter strip should have at least two of the following vegetation types:

- deep-rooted grasses, ground covers, or vines,
- deciduous and evergreen shrubs, or
- under- and over-story trees.

Native plant species should be used if possible. As newly constructed stormwater BMPs will be fully exposed for several years before the buffer vegetation becomes adequately established, plants that require full shade, or are susceptible to winter kill or prone to wind damage should be avoided and plant materials should conform to the American Standard for Nursery Stock (VA DCR, 1999). A schematic representation of a filter strip is given in Figure 3-6 (Claytor and Schueler, 1996).

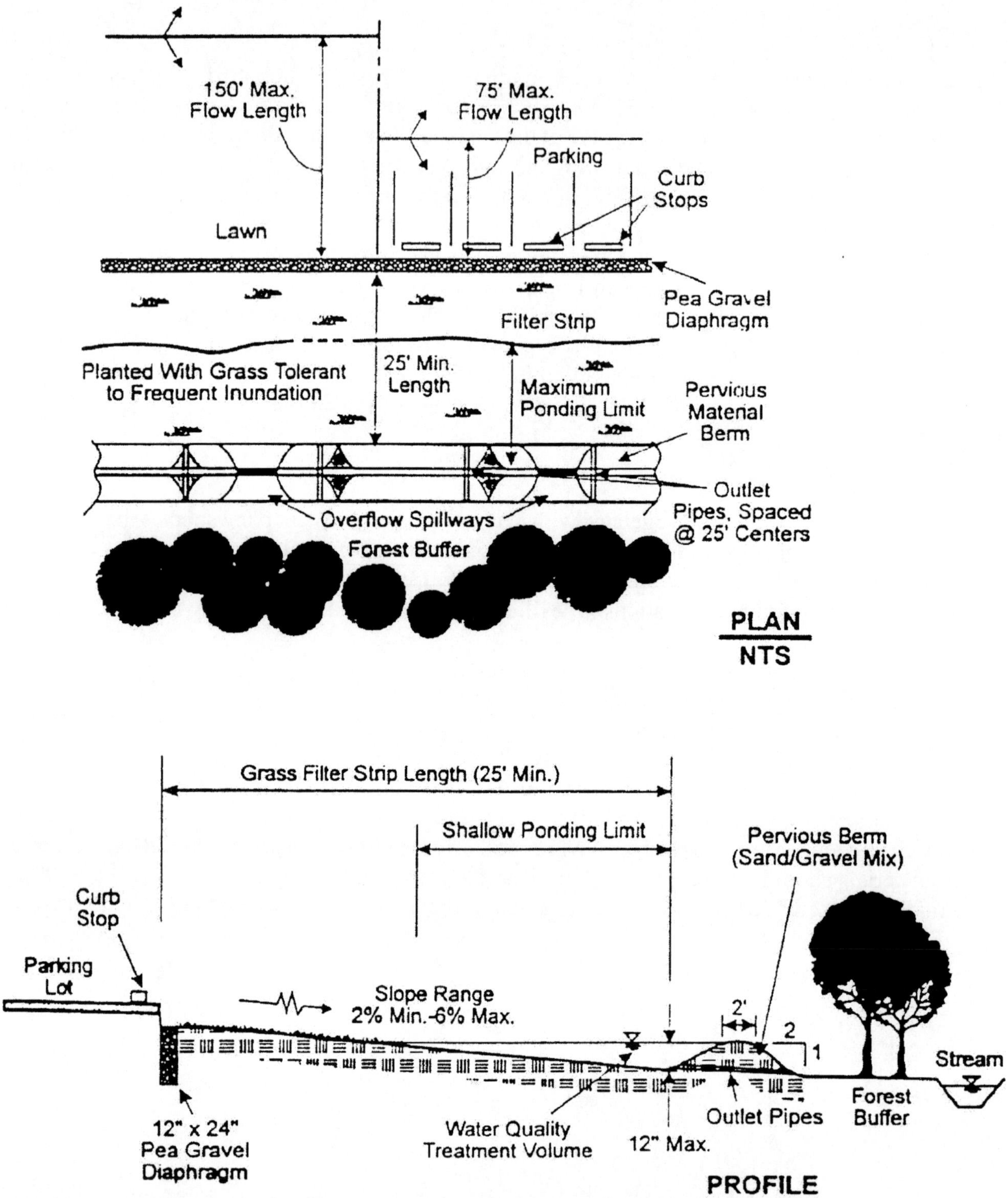

Figure 3-6. **Schematic of a Vegetated Filter Strip** (Claytor and Schueler, 1996)

3.3.6 Infiltration Trenches

A conventional infiltration trench is a shallow, excavated trench, generally 2- to 10- ft deep, that has been backfilled with a coarse stone aggregate and lined with filter fabric to create an underground reservoir for stormwater runoff from a specific design storm. Stormwater runoff diverted into the trench gradually infiltrates into the surrounding soils from the bottom and sides of the trench. This infiltration reduces the volume of runoff, removes many pollutants and provides stream baseflow and groundwater recharge. The design storm for an infiltration trench is typically a frequent, small storm such as the 1- yr event, that provides treatment for the "first flush" of stormwater runoff (0.5 in. runoff per acre of impervious surface) or even larger volumes.

The trench can be either an open surface trench or an underground facility (VA DCR, 1999) and infiltration trenches are typically implemented at the ground surface to intercept overland flows and stormwater runoff that generally enters the facility at one or more point sources. Primarily used as water quality BMPs, infiltration trench BMPs can route stormwater runoff into the aggregate filled storage chamber by two means: dispersed input or concentrated input. In dispersed input, water enters the top of the trench as overland sheet flow directed over a gently sloping grassed filter strip and flows to the surface of the storage chamber; concentrated input transports collected runoff to the storage chamber by means of gutters, curb inlets, and pipes. Some infiltration trench designs combine stormwater detention and water quality objectives by storing the entire stormwater volume with the water quality volume committed to infiltration by slowly releasing the water quality volume through an orifice set at a specified level in the storage reservoir (NVPDC and ESI, 1992).

Infiltration trenches require pretreatment of stormwater in order to remove as much of the suspended solids from the runoff as possible before it enters the trench. Also, public education with respect to street/driveway sediments may be provided in areas where an infiltration trench is proposed (Metro Council, 2001). Enhanced infiltration trenches have extensive pretreatment systems to remove sediment and oil. Conventional as well as enhanced trenches require on-site geotechnical investigations to determine appropriate design and location. Generally suited for low-to medium-density residential and commercial developments, these facilities can be incorporated in multi-use areas such as along parking lot perimeters, parking lots, residential areas, commercial areas, and open space areas. Unlike most BMPs, trenches can easily fit into the margin, perimeter, or other unused areas of developed sites, making them particularly suitable for retrofitting into existing developments or in conjunction with other BMPs (VA DCR, 1999). Infiltration may be a more promising practice in that it tends to reverse the hydrologic consequences of urban development by reducing peak discharges and increasing baseflow to local streams.

3.3.6.1 Stormwater Control

The size of the infiltration trench is determined by the volume of runoff controlled and the degree to which infiltration is used to dispose of runoff. There are three basic trench systems.

- Full exfiltration system - Runoff exits the stone reservoir by exfiltration through the underlying subsoil and the exfiltration system provides total peak discharge, volume and water quality control for all rainfall events less than or equal to the design storm (Schueler,

1987). The stone reservoir must be large enough to accommodate the entire increase in runoff volume for the design storm, less any runoff volume that is exfiltrated during a storm. Excess runoff from storms greater than the design storm should be handled by an emergency overflow channel such as a raised curb located above ground.

- Partial exfiltration system - In this design, an underground drainage system is installed that comprises regularly spaced perforated pipes located in shallow depressions to collect the runoff and direct it to a central outlet, and is generally designed to pass the 2-yr storm. Runoff from smaller storms will still be exfiltrated before it is collected, thereby providing significant water quality control. An alternative method may be to place perforated pipes on the underside near the top of the stone reservoir (NVPDC, 1987) to promote a greater degree of exfiltration, especially for design storms.

- Water quality exfiltration system - The storage volume of the stone reservoir is generally set to handle only the first flush of runoff volume during a storm, which has been variously defined as 0.5 in. of runoff per contributing impervious acre, 0.5 in. runoff per contributing total acres, and the volume of runoff produced by a 1 in. storm. Runoff volumes in excess of the first flush are not treated by the system but instead are conveyed to a stormwater management facility further downstream. While this system does not satisfy stormwater storage requirements, it may result in smaller, less costly facilities downstream (Schueler, 1987) .

3.3.6.2 Pollutant Removal Capability

The pollution removal processes that occur in infiltration systems are more complex than those occurring in wet ponds and extended detention dry ponds. Target pollutant behavior is governed by an array of factors including pH, redox potential, clay mineralogy, organic matter, microbial populations and temperature, as well as the physical characteristics of the soil environment, which change with depth and lateral distance inside the trench. While infiltration trenches are not really intended to remove a high level of coarse particulate pollutants, which need to be removed by a pre-treatment device before they enter the trench, fine particulates and soluble pollutants are effectively removed after exfiltrating through the trench and into the soil (MWCOG, 1979). Pollutant removal occurs due to sorption, precipitation, trapping, straining and bacterial degradation and transformation. It should be noted that the pollutant removal capability of water quality trenches are somewhat lower than other designs as a significant portion of the annual runoff volume will bypass a water quality trench, and is not subject to removal by exfiltration.

The pollution removal system of an infiltration system has two separate mechanisms. The sediment control system needed to maintain the function of the trench removes those pollutants associated with suspended solids such as adsorbed phosphorus, certain heavy metals and some exchangeable ions. Upon infiltration into the soil, several chemical and biological processes attenuate the levels of an array of pollutant species (NVPDC and ESI, 1992).

Infiltration trenches eliminate the discharge of the water quality volume to surface receiving waters and consequently can be considered to have 100% removal of all pollutants within this volume. Transport of some of these constituents to groundwater is likely, although the attenuation in the soil and subsurface layers will be substantial for many constituents (CASQA, 2003). The greatest sorption of nutrients and metals occurs in soils with a high content of clay and/or organic matter, with the least sorption observed in sandy soils (U.S. EPA, 1977); the same trend holds true for bacterial densities.

Positive factors influencing pollutant removal include:
- bank run or washed aggregate,
- high organic matter and loam content of subsoil,
- capture of a large fraction of annual runoff volume,
- effective pretreatment system, e.g., a sump pit; and
- pretreatment of sediments; oil; and grease.

Negative factors influencing pollutant removal include:
- sandy soils,
- trench clogging,
- high water table,
- long de-watering times,
- design considerations, and
- infiltration trench design variations.

3.3.6.3 Design Considerations

Trench designs can be distinguished as to whether they are located on the surface or below ground. Surface trenches accept diffuse runoff (sheet flow) directly from adjacent areas after it has been filtered through a grass buffer. Underground trenches accept more concentrated runoff (from pipes and storm drains), but require the installation of special inlets to prevent coarse sediment and oil/grease from clogging the stone reservoir.

Surface trenches are typically applied in residential areas, where smaller loads of sediment and oil can effectively be trapped by grass filter strips. As the surface is exposed, these trenches have a slightly higher risk of clogging than underground trenches, which could be prevented by placing a permeable filter fabric 6 to 12 in. below the surface of the trench for sediment interception. The following are a few design variations of the surface trench system.
- Median strip trench design
- Parking lot perimeter trench design
- Swale design

Underground trenches can be applied in a variety of development situations and are particularly suited to accept concentrated runoff. Pretreatment and the even distribution of concentrated runoff is an essential requirement in these systems. The top of the trench is protected by a layer of impermeable geo-textile and is covered by topsoil and planted with grass. Underground trenches may be more aesthetically pleasing, but may also be more expensive to maintain, and, more so when the trench is covered by pavement or concrete. These BMPs should only be installed when strong, enforceable maintenance agreements can be secured from the property owner. Some design variations include:

- over-sized pipe trench;
- underground trench with oil/grit inlet;
- under-the-swale design;
- dry well design; and,
- off-line trench system design.

General Considerations

Infiltration can be a very desirable method of stormwater treatment for land uses that do not heavily pollute stormwater runoff. It may be used where the subsoil is sufficiently permeable to provide a reasonable infiltration rate and where the water table is low enough to prevent pollution of groundwater. Areas containing karst topography may initially appear to have excellent infiltration, but are not recommended for planning an infiltration trench as they may cause subsurface collapse and sink-hole formation (VA DCR, 1999; Knoxville, 2003).

Paved areas subject to heavy use by motor vehicles, fueling stations, vehicle maintenance facilities, and similar areas subject to high hydrocarbon loads should be serviced by a water quality inlet as an in-line pretreatment to any infiltration structure (NVPDC and ESI, 1992).

Infiltration facilities are prone to high failure rates when designed improperly (Schueler, 1992b). This makes a strong case for designing and accepting infiltration trench systems on the basis of actual subsurface analysis and permeability tests rather than using pre-existing information on soils compiled from an array of data (VA DCR, 1999). Further, site-specific soil bores should be used to justify the use of infiltration practices. A minimum of one soil boring log is recommended for every 50 ft of trench length, with a minimum requirement of two soil boring logs for each proposed trench location (Metro Council, 2001). To identify localized soil conditions, soil boring should be done at the actual location of the proposed infiltration trench. In general, the following information should be included in a site-specific subsurface or geotechnical study.

Siting

One of the first steps in siting and designing infiltration treatment facilities is to conduct a characterization study. Geotechnical investigation data can be used for site characterization. Some of the key data and issues that need characterization include:

- surface features characterization,
- subsurface characterization,
- infiltration rate determination,
- soil testing, and
- infiltration receptor (WA DOE, 2001).

Soil Permeability

The soil types within the subsoil profile which extends a minimum of 3 ft below the bottom of the facility should be identified to verify the infiltration rate or permeability of the soil. The infiltration rate, or permeability, measured in in./h, is the rate at which water passes through the soil profile during saturated conditions, the minimum and maximum of which establish the suitability of various soil textural classes for infiltration. Each soil texture and the corresponding hydrologic properties within the soil profile are identified through analysis of a gradation test of the soil boring material. Soil textures acceptable for use with infiltration systems include those with infiltration rates between 0.52 in./h and 8.27 in./h (VA DCR, 1999), although Schueler (Schueler, 1987) recommends a minimum infiltration rate of 0.27 in./h (Table 3-7). This implies that sites with "D" soils (infiltration rates of less than 0.27 in./h), or any soil with a clay content greater than 30% (as determined from the SCS soil textural triangle) are not suitable options for infiltration trenches, nor are soils with a combined silt/clay percentage greater than 40% by weight that are susceptible to frost-heave. Silt loams and sandy clay loams ("C") soils provide marginal infiltration rates, and should only be considered for partial exfiltration systems. The stone subgrade must extend below the frost-line irrespective of the soil type, and is typically 8 to 12 in. in the Washington DC Metropolitan area. Also, trenches should not be located over fill soils that form an unstable upgrade and are prone to slope failure.

Under suitable soil conditions, soil cores or trenches to a depth of at least 5 ft below the anticipated level of the stone reservoir bottom may need to be evaluated for any impermeability in the soil strata that could impede infiltration. However, the presence of such layers does not necessarily preclude a trench, as long as the stone reservoir completely penetrates them.

Table 3-7. Soil Limitations for Infiltration Trenches

Soil Texture	Effective Water Capacity (C_w) in. /in.	Minimum Infiltration Rate (*f*) (in./h)	SCS soil group	Maximum Depth of Trench (in.) 48 h	72 h
*Sand	0.35	8.27	A	992	1489
**Loamy Sand	0.31	2.41	A	290	434
**Sandy Loam	0.25	1.02	B	122	183
**Loam	0.19	0.52	B	62	93
Silt Loam	0.17	0.27	C	32	49
Sandy Clay Loam	0.14	0.17	C	20	31

Clay Loam	0.14	0.09	D	11	16
Silty Clay Loam	0.11	0.06	D	7	11
Sandy Clay	0.09	0.05	D	6	9
Silty Clay	0.09	0.04	D	6	7
Clay	0.08	0.02	D	2	4

* Suitable for infiltration with typical 6' to 8' separation from seasonal high groundwater
** Suitable for infiltration with at least 3' separation from seasonal high groundwater
(Schueler, 1987; VA DCR, 1999; Knoxville, 2003)

Depth to Bedrock, Water Table, or Impermeable Layer

Typically, infiltration facilities are not recommended in areas with a high groundwater table due to the inability of the soil to adequately filter out pollutants before the stormwater enters the table. While the general requirement of various states is a distance of 2 to 4 ft, the Washington State Department of Ecology (WA DOE, 2001) recommends that the base of all infiltration trench systems shall be ≥5 ft above the seasonal high-water mark, bedrock (hardpan) or other low permeability layer. A minimum separation of 3 ft may be considered if the groundwater mounding analysis, volumetric receptor capacity, and the design of the overflow and/or bypass structures are judged to be adequate to prevent overtopping and meet the site suitability criteria (WY DEQ, 1999).

Topography

The topographic conditions of a development site represent feasibility factors that should be examined before designing an infiltration system. These factors include the slope of the land, the nature of the soil (natural/fill), and the proximity of building foundations and water supply wells. Infiltration trenches should be located in areas in which the slope does not exceed 20% (5H:1V) because steeper grade would increase the chance of water seepage from the subgrade to the lower areas of the site and reduce the volume that infiltrates. The use of infiltration trenches on fill material is not recommended due to the possibility of creating an unstable upgrade. Fill areas can be very susceptible to slope failure due to slippage along the interface of the in-situ and fill material, which could be aggravated if the fill material is allowed to become saturated by using infiltration practices (VA DCR, 1999).

Setback requirements for infiltration trenches that are required by local regulations, uniform building code requirements, or state regulations generally include the following:

- >100 ft from drinking water wells, septic tanks or drain fields, and springs used for public drinking water supplies. Infiltration trenches up gradient of drinking water supplies and within 1, 5 and 10-yr time of travel zones must comply with Health Department requirements,
- >20 ft downslope and >/100 ft upslope from building foundations,
- >20 ft from a Native Growth Protection and Easement (NGPE), and
- >50 ft from the top of slopes > 15% (WA DOE, 2001).

On-site and off-site structural stability due to extended subgrade saturation and/or head loading of the permeable layer need to be evaluated. This would include studying the potential impacts to downgradient properties, especially on hills with known side-hill seeps.

Design Criteria

Infiltration trenches are assumed to have rectangular cross-sections. Thus, the infiltration surface area (trench bottom) can be readily calculated from the trench geometry.

Sizing Procedure

The storage volume required for infiltration facilities designed for water quality enhancement is determined by the water quality volume, determined by the desired pollutant removal efficiency, and needs to be calculated using the void ratio of the backfill material that will be placed in it.

The sizing of water quality infiltration BMPs is best approached by applying Darcy's Law, which assumes that the drain time of the facility is controlled by one-dimensional flow through the bottom surface (VA DCR, 1999).

$$Q = f \times I \times SA \qquad \textbf{(3-9)}$$

where
Q = rate of exfiltration into soil ft^3/s;
f = infiltration rate of the soil in ft/h;
I = hydraulic gradient; and
SA = bottom surface area of facility in ft^2.

Infiltration Rate

Infiltration rates for treatment can be determined using either a correlation to grain size distribution from soil samples, textural analysis, or by in-situ field measurements. Short-term infiltration rates up to 2.4 in./h represent soils that typically have sufficient treatment properties, while long-term infiltration rates are used for sizing the trench based on maximum pond level and drawdown time. Long-term infiltration rates up to 2.0 in./h can also be considered for treatment if site suitability criteria are met for soil infiltration rate/drawdown time as well as soil physical and chemical suitability for treatment (WA DOE, 2001).

Historically, infiltration rates have been estimated from soil grain size distribution data using the USDA textural analysis approach. This involves conducting the grain size distribution test on soils passing the # 10 sieve (2 mm) (U. S. Standard) to determine the percentages of sand, silt and clay. The ASTM soil size distribution test procedure (ASTM D422), which considers the full range of soil particle sizes, is also being used by many laboratories to develop soil size distribution curves;

however, these should not be used in conjunction with the USDA soil textural triangle (WA DOE, 2001).

The three methods for determining the long-term infiltration rate for sizing the infiltration trench are:

- USDA soil textural classification,
- ASTM gradation testing at full scale infiltration facilities, and
- in-situ infiltration measurements or pilot infiltration tests (PIT) (WEF and ASCE, 1998).

Over the life of the infiltration facility, the rate of infiltration into the soil, f, may gradually decrease due to clogging of the surface layer of the soil as a result of siltation and biomass buildup in the trench. This suggests the need for a safety or a correction factor to be built into the design of the facility to allow for future clogging, which is a factor of 2 to be applied to the infiltration rate determined from the soil analysis. The design soil infiltration rate, f_d, therefore, is equal to one-half the actual rate:

$$f_d = 0.5 f \qquad \textbf{(3-10)}$$

It must be mentioned that a value of 2 for correction factor is based on the assumption that homogeneous soils should be used for treatment soil suitability determinations (WEF and ASCE, 1998), although a value between 2 and 4 but never less than 2, could be assigned, depending on the soil textural classification (WA DOE, 1991). These correction factors consider an average degree of long-term facility maintenance, TSS reduction through pretreatment, and site variability in the subsurface conditions that affect homogeneity. However, these correction factors could be reduced, subject to the approval of the local jurisdiction, under the following conditions for sites with little soil variability:

- where there will be a high degree of long-term facility maintenance, and
- where specific, reliable pretreatment is employed to reduce TSS entering the infiltration facility.

Correction factors higher than the general recommended values should be considered under the following situations:

- difficulty in implementing long-term maintenance,
- little or no pretreatment, and
- highly variable or uncertain site conditions (WA DOE, 1991).

Hydraulic Gradient

In areas with a shallow water table or impermeable layer, the hydraulic gradient may have an impact on the allowable design depth. The hydraulic gradient is given by equation 3-11 (VA DCR, 1999):

$$I = \frac{(h+L)}{L} \qquad \textbf{(3-11)}$$

where
I = hydraulic gradient;
h = height of the water column over the infiltrating surface (ft); and
L = distance from the top surface of the BMP to the water table, bedrock, impermeable layer, or other soil layer of a different infiltration rate (ft).

The hydraulic gradient will be assumed to be equal to one in all infiltration designs since the gradient approaches unity as the facility drains. Therefore,

$I = 1$

Maximum Ponding or Storage Time and Trench Depth

The minimum and maximum time for the trench to empty the stormwater volume into the soil by infiltration is based upon balancing optimum pollutant removal and assuring adequate stormwater management performance. Trenches should be designed in general to provide a detention time of 6 to 72 h. A minimum drainage time of 6 h should be provided to ensure satisfactory pollutant removal in the infiltration trench (Schueler, 1987). Trenches may be designed to provide temporary storage of stormwater, yet it should drain prior to the next storm event. The drainage time will vary by precipitation zone and the maximum drain time for the total design infiltration volume varies from 24 (WA DOE, 2001) to 72 h (Metro Council, 2001). The Northern Virginia Planning District Commission (NVPDC and ESI, 1992) recommends that the infiltration trench be designed with a maximum of 48 h for the water quality volume, 72 h for the total volume, and with a minimum retention time of 24 h for the water quality volume. According to the Commonwealth of Virginia Department of Conservation and Recreation (VA DCR, 1999), following the occurrence of a storm event, all infiltration trenches should be designed with a maximum drain time, T_{max}, of 48 h for the water quality volume. The maximum drain time, along with the minimum design soil infiltration rate, f_d, as verified through a subsurface investigation and analysis, will dictate the maximum allowable design depth, d_{max}, of the structure. The maximum depth for an infiltration trench may be defined as:

$$d_{max} = \frac{f_d T_{max}}{V_r} \quad \textbf{(3-12)}$$

where
d_{max} = maximum allowable depth of the trench, in ft;
f_d = design infiltration rate of the trench area soils, in ft/h ($f_d = 0.5f$);
T_{max} = maximum allowable drain time (48 h)
V_r = void ratio of the stone reservoir expressed in terms of the percentage of porosity divided by 100 (0.4 typically)
A void ratio of 0.40 is assumed for stone reservoirs using 1.5 to 3.5 in. stone - VDOT No. 1 Coarse-graded Aggregate (VA DCR, 1999).

The minimum surface area of the facility bottom may be defined by equation 3-13 :

$$SA_{\min} = \frac{Vol_{wq}}{f_d T_{\max}} \tag{3-13}$$

where
$SA_{\min}$ = minimum trench bottom surface area, in ft^2;
Vol_{wq} = water quality volume requirements, in ft^3;
f_d = design infiltration rate of the trench area soils, in ft/h ($f_d = 0.5f$)
$T_{\max}$ = maximum allowable drain time = 48 h

The storage volume of the facility is defined as: $L \times W \times D \times V_r$ **(3-14)**

Determination of the dimensions of the storage reservoir is made by fitting the length, width and depth into a configuration that satisfies drain-time and storage volume requirements while keeping the storage reservoir bottom within the optimum depth for infiltration (NVPDC and ESI, 1992). A long, narrow trench is less affected by water table mounding and is advisable when the depth to seasonal high water table or bedrock is within 5 ft of the trench bottom. In order to keep the trench bottom elevation within the optimum depth in the soil profile, long trenches may need to be curved parallel to the topographic contour. If greater storage is needed than the design storm volume requirement, the trench dimensions could be adjusted by the following recommendations, by order of priority:

- Increase the length of the trench if the seasonal high water table or bedrock is within 5 ft of the trench bottom.
- Increase the width, if the length cannot be increased due to site constraints.
- It is permissible to increase the depth if the seasonal high water table and bedrock are known to be at a depth greater than 5 ft below the bottom of the trench, provided that the new bottom elevation meets the same criteria for optimum depth.

Most infiltration trenches are generally greater than 2 ft in depth, but the frost depth needs to be considered in shallow design trenches. The bottom of the structure should be 18 in. below the surface to avoid freezing of the trench bottom surface.

Backfill Material

Backfill material for the trench should be clean aggregate with a maximum diameter of 3.5 in. and a minimum diameter of 1.5 in., and the aggregate should contain few aggregates smaller than the selected size. An 8 in. bottom sand layer is required for most of the trenches to promote better drainage and reduce the risk of soil compaction when the trench is backfilled with stone (Schueler, 1992b).

Filter Fabric

The sides and bottom of the trench should be lined with geotextile fabric (filter fabric). For an aggregate surface trench, filter fabric should surround all of the aggregate fill material except the top one foot (VA DCR, 1999). A separate piece of fabric should be used for the top layer to act as a failure plan. There can be a layer of non-woven filter fabric 6 to 12 in. below the ground surface to prevent suspended solids from clogging the majority of the storage media. The filter fabric may need frequent replacement, depending on the volume of suspended solids transported to the trench.

The filter fabric material must be compatible with the surrounding soil textures and application purposes, with the cut width of the filter fabric having sufficient material for a minimum 12 in. overlap. When overlaps are required between rolls, the upstream roll must lap a minimum of two ft over the downstream roll to provide a shingled effect. The bottom of the infiltration trench can be covered with a 6 to 12 in. layer of clean sand in place of filter fabric.

Storage Media

The basic infiltration trench design utilizes stone aggregate in the top of the trench to provide adequate void space (at least 40%) (Schueler, 1987) for filtering and removing pollutants. The trench should be filled with clean, washed stone with a diameter of 1.5 to 3 in. Pea gravel could also be substituted for stone aggregate in the top 0.3 meter (1 ft) of the trench, as it improves sediment filtering and maximizes pollutant removal in the top of the trench. When these modified trenches become clogged, they can generally be restored to full performance by removing and replacing only the pea gravel layer, without replacing the lower stone aggregate layers.

Observation Well

An observation well should be installed for every 50 ft of infiltration trench length. The purpose of the well is to show how quickly the trench dewaters following a storm, as well as providing a means of determining when the filter fabric is clogged and requires maintenance. It should be installed in the center of the structure, flush with the ground elevation of the trench. This can be a 4 to 6 in. diameter PVC pipe, anchored vertically to a foot plate at the bottom of the trench, and the well should have a lockable above-ground cap (Metro Council, 2001).

Overflow Channel

Although an emergency spillway is not necessary because of the small drainage areas controlled by an infiltration trench, the overland flow path taken by surface runoff when the trench capacity is exceeded needs to be evaluated. A non-erosive overflow channel leading to a stabilized water course should be provided, as necessary, to insure that uncontrolled, erosive, concentrated flow does not develop.

Pretreatment

Infiltration trenches are susceptible to high failure rates due to clogging from sediments, and therefore require pretreatment of stormwater in order to remove as much of suspended solids as possible from the runoff before it enters the trench. Pretreatment such as grit chambers, swales with check dams, filter strips, or sediment forebays/traps should be a fundamental component of any BMP system relying on infiltration. Pretreatment facilities should be installed off-line in order to reduce both the frequency of turbulent flow-through and the associated scour and/or resuspension of residual material (Schueler, 1992b).

A grass strip or other type of vegetated buffer at least 20 ft wide should be maintained around trenches that accept surface runoff as sheet flow. The slope of the filter strip should be approximately 1% along its entire length and 0% across its width. A minimum filter length of 50 ft is desirable for areas receiving high loads of suspended solids.

All trenches with surface inlets should be engineered to capture sediment from the runoff before it enters the stone reservoir. The design of the trench must include a pretreatment facility design, complete with maintenance and inspection requirements.

Bypass

A bypass system should be implemented for all infiltration trenches. A bypass flow path should be incorporated in the design of an infiltration trench to convey high flows around the trench. The overland flow path of surface runoff exceeding the capacity of the infiltration trench should be evaluated to preclude erosive concentrated flow. If computed flow velocities do not exceed the non-erosive threshold, overflow may be accommodated by natural topography.

Groundwater Mounding

Groundwater mounding means the local elevation of the water table as a result of infiltrated surface water, and calculations may be necessary in cases where slope stability is a concern, and/or a high water table is encountered. The results from these calculations should be regarded as an indication of the mounding potential rather than as an accurate representation of the actual mounding depth. Figure 3-7 is an example schematic of an infiltration trench (MDE, 2000).

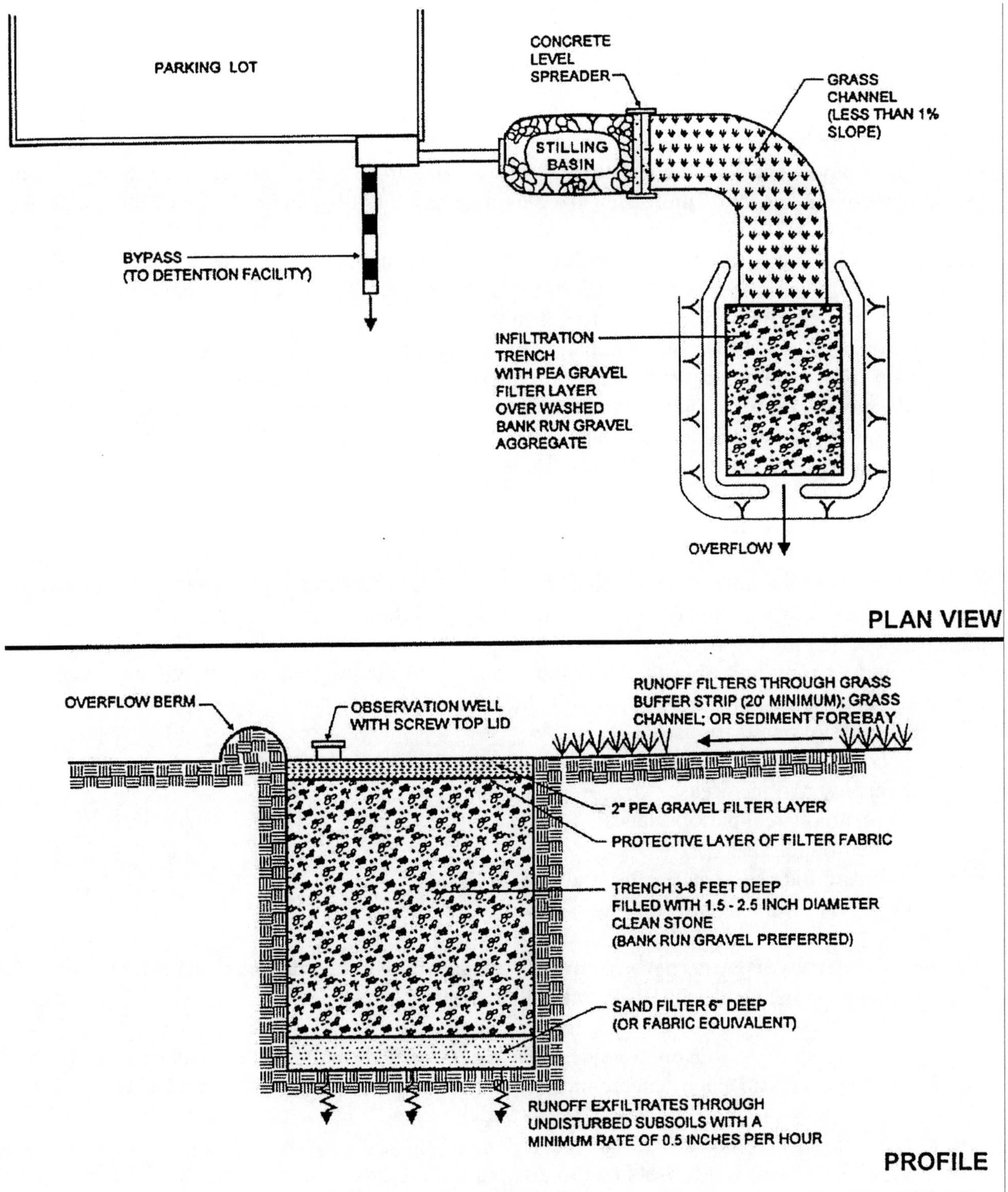

Figure 3-7. **Schematic of an Infiltration Trench** (MDE, 2000)

3.3.7 Porous Pavement

Porous concrete and asphalt pavements are being used as BMPs and replace conventional asphalt pavement or other hard paving surfaces whereby runoff is diverted through a porous asphalt layer and into an underground stone reservoir. The stored runoff then gradually infiltrates into the subsoil. The basic porous pavement system consists of a top layer of porous asphalt concrete covering a layer of gravel that covers a layer of uniformly sized aggregate, which is placed on top of the existing soil sub-base (Schueler, 1987). Stormwater penetrates the porous asphalt and is filtered through the first layer of gravel. The voids in the lower level of large aggregate are filled with runoff. The stored runoff gradually infiltrates into the underlying soil. A sheet of filter fabric below the aggregate prohibits the underlying soil from entering and clogging the facility (NVPDC and ESI, 1992). Provided that the grades, subsoil drainage characteristics, and groundwater table conditions are suitable for its use, porous pavement can be effectively used to recharge groundwater supplies and reduce stormwater runoff as well as water pollution from paved low volume traffic areas (WA DOE, 2001). When properly designed and carefully installed, porous pavement has load bearing strength, longevity, and maintenance requirements similar to conventional pavement.

The surface of the pavement is designed to provide adequate strength to accommodate vehicles while allowing infiltration of surface water and filtration of pollutants. If infiltration into the soil is not practical, the filtered runoff can be discharged through a sub-base drainage system that would outfall into a storm sewer system or a natural drainage path. Pollutant filtration is greatly reduced when the pavement drains into a storm sewer. Several studies have concluded that porous asphalt pavement is sufficiently strong and able to withstand freeze/thaw cycles such that it will last as long as conventional pavement. Porous pavement systems are typically used in low-traffic areas such as the following types of applications:

- parking pads in parking lots,
- overflow parking areas,
- residential street parking lanes,
- recreational trails,
- golf cart and pedestrian paths, and
- emergency vehicle and fire access lanes.

There are three types of porous pavement: porous asphalt pavement; porous concrete pavement; and, modular porous concrete block (WEF and ASCE, 1998).

Porous asphalt pavement is an open-graded coarse aggregate, bound together by asphalt cement into a coherent mass, with sufficient interconnected voids to provide a high rate of permeability to water.

Porous concrete (also referred to as enhanced porosity concrete, porous concrete, Portland cement and pervious pavement) is a subset of a broader family, including porous asphalt, and various kinds of grids and paver systems. Also known as "no fines concrete," it is a special type of concrete that allows stormwater to pass through it, thereby reducing the runoff from a site. In addition, porous concrete provides runoff treatment through filtration and allows for ground water recharge. Porous concrete or "no fines concrete paving" is a structural, open textured pervious concrete paving surface consisting of standard Portland cement, fly ash, locally available open graded coarse aggregate, admixtures, fibers, and potable water. When properly handled and installed, porous concrete has a high percentage of void space (approximately 17 to 22%) which allows rapid percolation of stormwater through the pavement. Porous concrete is thought to have a greater ability than porous asphalt to maintain its porosity in hot weather and thus is provided as a limited application control. Although, porous concrete has seen growing use in Georgia, there is still very limited practical experience with this measure. Porous concrete is designed primarily for stormwater quality, i.e., the removal of stormwater pollutants. However, they can provide limited runoff quantity control, particularly for smaller storm events. For some smaller sites, trenches can be designed to capture and infiltrate the channel protection volume (Cp_v) in addition to WQ_v. Porous concrete will need to be used in conjunction with another structural control to provide overbank and extreme flood protection, if required (Atlanta Regional Commission, 2001).

Modular porous pavers are structural units, such as concrete blocks, bricks, or reinforced plastic mats, with regularly inter-dispersed void areas used to create a load bearing pavement surface. The void areas are filled with pervious materials (gravel, sand, or grass turf) to create a system that allows for the infiltration of stormwater runoff. Porous paver systems provide water quality benefits in addition to groundwater recharge and a reduction in stormwater volume. The use of porous paver systems results in a reduction of the effective impervious area on a site. There are many different types of modular porous pavers available from different manufacturers, including both pre-cast and mold in-place concrete blocks, concrete grids, interlocking bricks, and plastic mats with hollow rings or hexagonal cells. Modular porous pavers are typically placed on a gravel (stone aggregate) base course. Runoff infiltrates through the porous paver surface into the gravel base course, which acts as a storage reservoir as it exfiltrates to the underlying soil. The infiltration rate of the soils in the subgrade must be adequate to support drawdown of the entire runoff capture volume within 24 to 48 h. Special care must be taken during construction to avoid undue compaction of the underlying soils, which could affect the soils' infiltration capability.

The construction of porous asphalt and concrete are similar to a conventional pavement, except that sand and finer fraction of the aggregate are left out of the pavement mix, and is typically placed on top of a granular base. The modular block pavement is constructed by placing the blocks over a layer of coarse gravel, which in turn is located on a porous geotextile fabric layer. Porous concrete and asphalt pavements have a tendency to clog and seal within 1 to 3 yr (Urbonas and Stahre, 1993), with faster sealing rates reported in areas with excessive winter salting and sanding. Notable exceptions to this were the concrete pavement installations in the state of Florida. Interlocking cellular concrete block pavement seems to seal at a slower rate and has a good record of service under a wide range of climatic conditions (WEF and ASCE, 1998).

3.3.7.1 Stormwater Control

Based on the runoff storage provided by the stone reservoir and the degree of reliance on exfiltration, porous pavement designs fall into three basic categories: complete exfiltration systems; partial exfiltration systems; water quality exfiltration systems (discussed in 3.3.6.).

3.3.7.2 Pollutant Removal Capability

Porous pavement systems in operation show high removal rates for sediment, nutrients, organic matter, and trace metals. The majority of the removal occurs as a result of the exfiltration of runoff into the subsoil, and subsequent adsorption or straining of pollutants within the subsoil (WY DEQ, 1999). Mechanisms of removal include adsorption, straining, and microbial decomposition in the subsoil below the aggregate chamber, and trapping of particulate matter within the aggregate chamber. The first pollutant removal process occurs in the large aggregate reservoir wherein pollutants adsorb to and are absorbed by the aggregate material. Suspended matter will settle out at the bottom of the aggregate layer. The second process for removing pollutants occurs only if the runoff drains into the soil instead of being discharged by a drain. Pollutants that enter the soil sub-base are also adsorbed to and absorbed by the soil particles in addition to aerobic decomposition as well as chemical precipitation of the pollutants within the soil strata (NVPDC and ESI, 1992).

Positive factors influencing pollutant removal include:

- high exfiltration volumes,
- high surface area,
- routine vacuum sweeping,
- maximum drainage time two days,
- highly permeable soils,
- clean-washed aggregate,
- organic matter in subsoils, and
- pre-treatment of off-site runoff.

Negative factors influencing pollutant removal include:

- poor construction practices,
- inadequate surface maintenance,
- use of sand during snow conditions, and
- low exfiltration volumes (WY DEQ, 1999).

3.3.7.3 Design Considerations

Siting

A prerequisite in the construction of porous pavement systems is the evaluation of the site for feasibility to rely on exfiltration to dispose of runoff. The use of porous pavement is highly constrained, requiring deep and permeable soils, restricted traffic, and suitable adjacent land uses Use may also be restricted in regions with colder climates, arid regions, or regions with high wind erosion rates and in areas of sole-source aquifers (WY DEQ, 1999). Pretreatment using filter strips or vegetated swales for removal of coarse sediments is recommended (Atlanta Regional Commission, 2001).

As porous pavements cannot withstand the passage of heavy trucks due to a lower tensile strength than a conventional pavement, these are typically recommended for lightly used satellite parking areas and access roads.

The design of porous pavement systems should include a seepage analysis. Possible adverse impacts of seepage from infiltration measures to building foundations, basements, roads, parking lots, and sloping areas should be addressed. It is recommended that the porous pavement be located 10 or more ft down gradient of foundation walls, particularly in residential areas (NVPDC and ESI, 1992).

Porous pavement systems should be located at least 100 ft away from a drinking water well to minimize the possibility of groundwater contamination, at least 10 ft down-gradient from nearby building foundations, and at least 100 ft up-gradient.

Porous concrete systems should typically be used in applications where the pavement receives tributary runoff only from impervious areas. If runoff is coming from adjacent pervious areas, it is important that those areas be fully stabilized to reduce sediment loads and prevent clogging of the porous paver surface. Any significant amount of offsite flow should be diverted around the pavement surface. Limited offsite runoff and all onsite runoff should be filtered before it flows over the pavement.

To protect groundwater from potential contamination, runoff from designated hotspot land uses or activities must not be infiltrated. Porous concrete should not be used under the following conditions:

- manufacturing and industrial sites, where there is a potential for high concentrations of soluble pollutants and heavy metals;
- areas with a high pesticide concentration; and
- areas with karst geology without adequate geotechnical testing by qualified individuals and in accordance with local requirements (Atlanta Regional Commission, 2001).

Soils

Porous pavement is not suitable for sites with soil infiltration rates of less than 0.5 in./h (D soils), or any soils with a clay content greater than 30%. C soils (silt loam and sandy clay loams) provide marginal infiltration rates, and should probably only be considered for partial exfiltration systems (Schueler, 1987; Atlanta Regional Commission, 2001). Soils with a combined silt/clay content of over 40% by weight are susceptible for frost heave, and may not be suited for these applications. These systems should never be constructed over fill soils, which often form an unstable upgrade, and are prone to slope failure. Stone subgrade must extend below the frost line irrespective of soil conditions. During construction and preparation of the subgrade, special care must be taken to avoid compaction of the soils.

The most critical factor in determining the applicability of porous pavement as a BMP device is the infiltration capacity of the underlying soil (NVPDC and ESI, 1992). Core samples or trenches at least 2 to 4 ft below the anticipated level of the bottom of the stone reservoir should be examined for any impermeable soil strata that might impede infiltration, such as localized clay lenses, hardpans, or fragipans. Subsurface drainage may be required if the soil does not exhibit adequate infiltration capacity. Subsoils are generally susceptible to frost heave if the soil contains more than 3% of particles smaller than 0.02 mm in diameter. Such soils do not allow the infiltration from the facility and should be avoided (NVPDC and ESI, 1992).

Slope

Porous concrete systems should not be used on slopes greater than 5%; 2% grade is recommended. For slopes greater than 1%, barriers perpendicular to the direction of drainage should be installed in sub-grade material to keep it from washing away, or filter fabric should be placed at the bottom and sides of the aggregate to keep soil from migrating into the aggregate and reducing porosity.

Depth to Bedrock and Seasonally High Water Table

The depth from the bottom of the gravel base course to the level of the seasonally high water table or to bedrock must be sufficient (2 to 4 ft) to allow for adequate infiltration and filtering of water released through the bottom of the structure. A minimum of 3 ft (preferably 4 ft) of clearance is needed between the bottom of the stone reservoir and the bedrock level. This data can be inferred from local soil data maps, but needs to be confirmed by actual soil test bores (WY DEQ, 1999). To insure complete draining of the stone reservoir in 72 h, it may be necessary to limit the depth of the stone reservoir if underlying soils have relatively low exfiltration rates. Soil limitations for porous pavement are shown in Table 3-8.

Table 3-8. Soil Limitations for Porous Pavement

Soil Type	**Minimum Infiltration Rate (*f*) (in./h)**	**SCS Soil Group***	**Maximum Depth of Storage** (in.)**	
			48 h	**72 h**
Sand	8.27	A	992	595
Loamy Sand	2.41	A	290	174
Sandy Loam	1.02	B	122	183
Loam	0.52	B	62	93
Silt Loam	0.27	C	32	49

* Sandy Clay Loams, Clay Loams, Silty Clay Loams, Sandy Clay, Silty Clay, and Clay soils are not included as these soil types are all not feasible for infiltration basins.

**Maximum Depth of stone reservoir that can drain completely within 48 or 72 h after a storm, given the soil infiltration rate.

(Schueler, 1987)

Watershed Size

The most suitable drainage area for porous pavement sites should be restricted to between 0.25 and 10 acres. This guideline tends to reflect the perceived economic and liability problems associated with larger applications and the cost-effectiveness of other BMPs outside of this range.

Design Parameters

Since the surface area of the porous pavement will typically depend on how large a parking lot will be built, the critical design consideration will be the depth of the large aggregate layer. As with infiltration trenches, the maximum depth of the large aggregate layer is a function of allowable detention time, the porosity of the aggregate, and the soil infiltration rate. The bottom of the facility should be below the frost line and approximately 4 ft above bedrock and the level of the seasonally high water table. The same design steps presented for infiltration facilities earlier in this chapter apply to porous pavement, and include the additional step of determining the thickness of the porous pavement layer. The depth of the asphalt layer and underlying stone reservoir depends on the strength of the sub-base soil and the projected traffic intensities (NVPDC and ESI, 1992).

The following list of general design elements should be considered in any porous pavement design:

- anticipated traffic intensities, defined by the average daily equivalent axle load (EAL);
- California Bearing Ratio (CBR) of the soils; and
- susceptibility of the soils to forest heave.

Methods for conducting the CBR test are described in ASTM D1883 and AASH0 T193 (VA DCR, 1999). The asphalt layer is typically 2.5 to 4 in. thick. The minimum combined thickness of the asphalt layer and stone reservoir can be determined from the Table 3-9.

Table 3-9. Minimum Thickness of Porous Paving

Traffic Group	**General Character**	**California Bearing Ratio**				**EAL**
		>15	**10-14**	**6-9**	**<5***	
1	Light Traffic	5"	7"	9"		<5
2	Medium Light Traffic (Max. 1,000 VPD)	6"	8"	11"		6-20
3	Medium Traffic (Max. 3,000 VPD)	7"	9"	12"		21-75

*Studies indicate that for all traffic groups with CBR of 5 or less, the subgrade was improved to CBR 6 with crushed stone 2 in. size.
VPD = Vehicles Per Day
EAL = Equivalent Axle Load (18 Kips) average daily
Note: Thicknesses refer to the minimum combined depth of asphalt layer and stone reservoir necessary to carry appropriate load.
(NVPDC and ESI, 1992)

The following design procedure represents a generic list of the steps typically required for the design of porous pavement:

- Determine if the anticipated development conditions and drainage area are appropriate for a porous pavement application.
- Determine if the soils (permeability, bedrock, water table, Karst, etc.) and site topographic conditions (slopes, etc.) are appropriate for a porous pavement application.
- Locate the porous pavement section on a site within topographic constraints.
- Determine the drainage area for the porous pavement and calculate the required water quality volume.
- Evaluate the hydrology of the contributing drainage area to determine peak rates of runoff.
- Design the porous pavement stone reservoir; e.g., as shown in the Virginia BMP manual.
  - Design infiltration rate, $f_d = 0.5 f$
  - Max. storage time $T_{max} = 48$ h
  - Max. storage depth, d_{max}
  - Stone backfill of clean aggregate (1.5 to 3.5 in.) VDOT No. 1 open-graded coarse aggregate
  - Filter gravel layer - 2 in. of clean aggregate (0.5 in.) VDOT No. 57 open-graded coarse aggregate
  - Sand layer on trench bottom (8 in.) or filter fabric, per geotechnical and pavement design recommendations
  - Filter fabric on trench sides and top (not on trench bottom) keyed into trench
- Overflow channel or large storm bypass.
- Observation well.
- Provide pavement section design and material specifications.
- Provide sequence of construction.
- Provide maintenance and inspection requirements (VA DCR, 1999).

A schematic representation of a porous pavement is presented in Figure 3-8 (Schueler, 1987).

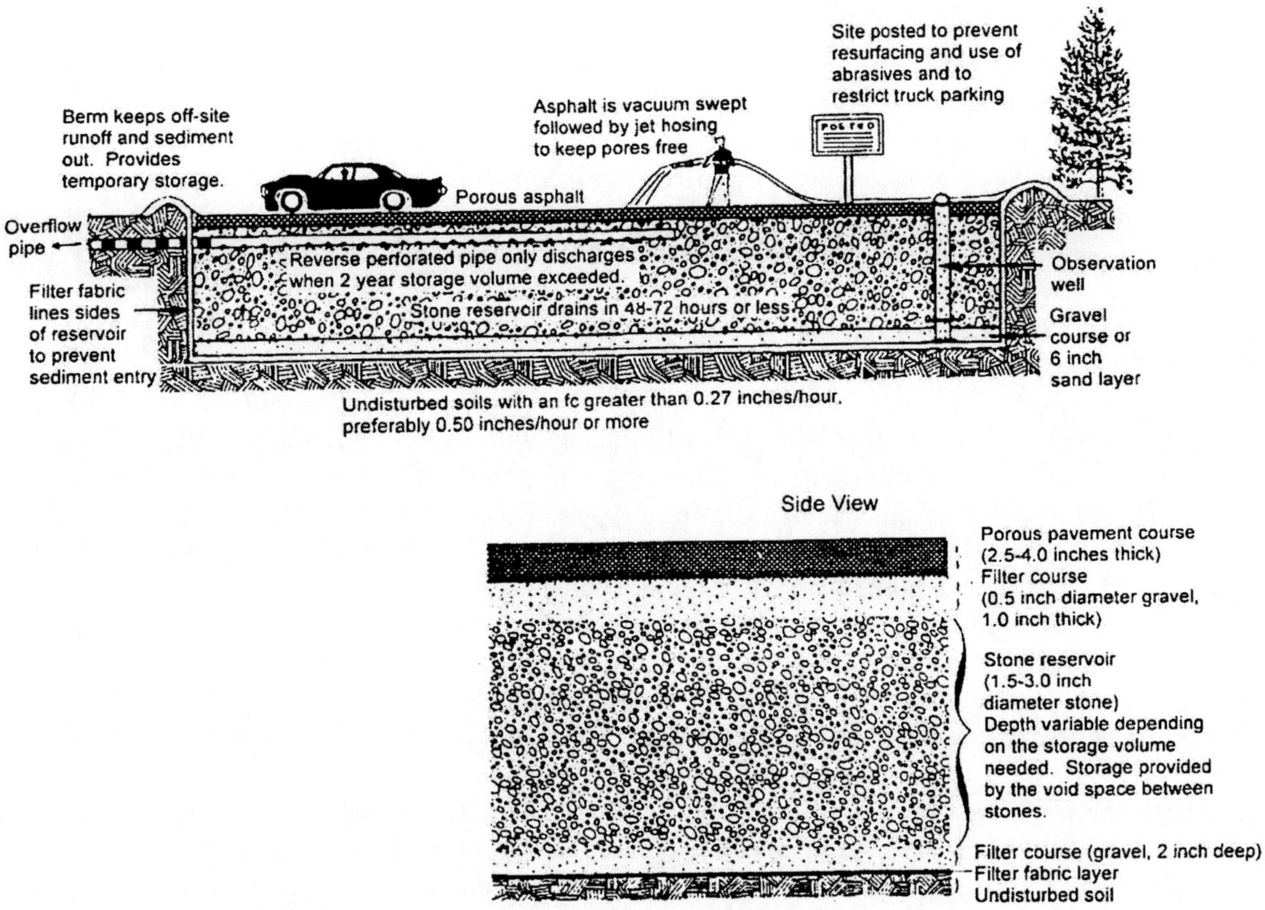

Figure 3-8. Schematic of a Porous Pavement (Schueler, 1987)

3.3.8 Sand and Organic Filters

Sand filters are structural stormwater controls that capture and temporarily store stormwater runoff and pass it through a filter bed of sand. They have been successfully used in Austin, TX, the District of Columbia, the state of Delaware, and in Alexandria, VA over the last two decades (VA DCR, 1999). Most sand filter systems consist of two-chamber structures. The first chamber is a sediment forebay or sedimentation chamber, which removes floatables and heavy sediments. The second, which is a filtration chamber, removes additional pollutants by filtering the runoff through a sand bed. The filtered runoff is typically collected and returned to the conveyance system, though it can also be partially or fully exfiltrated into the surrounding soil in areas with porous soils.

Sand filters may be "unconfined" sand-filled trenches with perforated underdrains or "confined" systems where the filter medium is contained in a concrete vault with a drain at the bottom of the vault. Depending on the specific design, these types of filters are often referred to as "Delaware Filters" or "Austin Filters" after the localities where they were originally designed and installed. Large sand filters are installed above ground and are self-contained sand beds that can treat stormwater from drainage areas as much as 5 acres in size. Enhanced sand filters utilize layers of peat, limestone, leaf compost, and/or topsoil, and may also have a grass cover crop. The adsorptive media of enhanced sand filters is expected to improve removal rates (WY DEQ, 1999). Sand filters can fall under two basic designs: (i) surface sand filter; and, (ii) perimeter sand filter.

The surface sand filter is a ground-level open air structure that consists of a pretreatment sediment forebay and a filter bed chamber. This system can treat drainage areas up to 10 acres in size and is typically located offline. Surface sand filters can be designed as an excavation with earthen embankments or as a concrete or block structure.

The perimeter sand filter is an enclosed filter system typically constructed just below grade in a vault along the edge of an impervious area such as a parking lot. The system consists of a sedimentation chamber and a sand bed filter. Runoff flows into the structure through a series of inlet grates located along the top of the system.

Yet another design variant, the underground sand filter, is intended primarily for extremely space-limited and high density areas, and is considered a limited structural application control (Atlanta Regional Commission, 2001).

3.3.8.1 Stormwater Control

Sand filter systems are designed primarily as offline systems for stormwater quality and typically need to be used in conjunction with another structural control to provide downstream channel protection, overbank flood protection, and extreme flood protection, if required. However, under certain circumstances, filters can provide limited runoff quantity control, particularly for smaller storm events.

3.3.8.2 Pollutant Removal Capability

Pollutant removal is primarily achieved by straining pollutants through the filtering medium (sand or peat) and settling on top of the sand bed and/or pretreatment pool. A grass cover crop on the filter helps in the additional removal of nutrients by plant uptake. Sand filter removal rates are high for sediment and trace metals, and moderate for nutrients, biochemical oxygen demand (BOD) and fecal coliform (FC) (City of Austin, 1991).

Positive factors influencing pollutant removal include:

- offline systems,
- peat and/or limestone layer,
- grass cover,
- longer draw down times ranging from 24 to 40 h,
- pretreatment pool,
- minimum depth of 18 in.,
- regular maintenance, and
- no direct connection to groundwater.

Negative factors influencing pollutant removal include:

- online systems, and
- freezing weather.

3.3.8.3 Design Considerations

General Considerations

Several types of intermittent sand filter facilities are recognized for stormwater quality management purposes, and the general design criteria presented below apply to the design of these facilities for water quality control. This implies that the volume of runoff to be treated is determined by the water quality volume and the desired pollutant removal efficiency (VA DCR, 1999).

The Austin, Texas Filter

The concept and use of surface filters initially originated in Austin, Texas, where these filters have been extensively used in catchments of up to 20 ha (50 acres). Austin filter has two design variants, one with full sedimentation and the other with partial sedimentation. The full sedimentation configuration includes a sedimentation basin designed to hold the entire water quality volume (i.e., equivalent to the 40 h drain time maximized volume) and to release this volume to the filter over a 40 h drawdown period. This system should be used unless topographical constraints make this design unfeasible. The partial sedimentation configuration requires less depth than the full sedimentation system and may be applicable where topographical constraints exist. In this system, a smaller sedimentation chamber is located upstream of the filtration basin, is designed to remove the heavier sediment and trash litter only, and requires more intensive maintenance than the full sedimentation system. The volume of the sediment chamber should be no less than 20% of the water quality volume used for the full sedimentation design. The design must ensure that the sediment chamber discharges the flow evenly(WEF and ASCE, 1998).

Linear Filter - Delaware

The Delaware Filter is an underground system that uses a vault with a permanent pool of water as the pretreatment device. Recommended for catchments of up to 2 ha (5 acres), the volume of both the sedimentation and filter chambers are approximated to 38 m^3/ha (540 ft^3 per contributing acre) and the surface area of each chamber should be 25 m^3/ha. Pavement and inlet design and construction are critical in a Delaware filter (the filter should be positioned relative to the pavement to evenly distributed the flow as it enters the sedimentation chamber) (WEF and ASCE, 1998).

Underground Vault - Washington D. C

The initial settling chamber is undersized for effective sedimentation, causing the filter to clog quickly. When the filter clogs, the flow simply overtops the overflow weir and flows directly to the outlet, with no indication that the filter is plugged. This filter type when used, should be sized using the Delaware linear filter criteria, including the pre-settlement chamber. It is also strongly recommended that the overflow weir and de-watering drain in the filter chamber be blocked and that the entrance manhole covers over the sedimentation chamber and the outflow chamber be replaced with grates. If the filter clogs, the water will back up in the vault, overflow out of the inlet grate over the sedimentation compartment, and back into the outfall chamber, giving a clear visual indication that the filter is plugged(WEF and ASCE, 1998).

Location and Siting

Surface sand filters should have a contributing drainage area of 10 acres or less. The maximum drainage area for a perimeter sand filter is 2 acres.

Sand filter systems are generally applied to land uses with a high percentage of impervious surfaces. Sites with less than 50% imperviousness or high clay/silt sediment loads must not use a sand filter without adequate pretreatment due to potential clogging and failure of the filter bed. Any disturbed areas within the sand filter facility drainage area should be identified and stabilized.
Surface sand filters are generally used in an offline configuration where the water quality volume (WQ_v) is diverted to the filter facility through the use of a flow diversion structure and flow splitter. The diversion structure or flow splitter is used to divert stormwater flows greater than the WQ_v to other controls or downstream. Perimeter filters are typically sited along the edge or perimeter of an impervious area such as a parking lot.

Sand filter systems are designed for intermittent flow and must be allowed to drain and re-aerate between rainfall events and should not be used on sites with a continuous flow from groundwater, sump pumps, or other sources.

Physical Specifications/Geometry

An access ramp with a slope not exceeding 7:1 or the equivalent should be included at the inlet and outlet of a surface filter for maintenance purposes. Side slopes for earthen or grass embankments should not exceed 3:1 (H:V) to facilitate mowing and site slope should be no more than 6% across filter location (Metro Council, 2001).

A major drawback for a media filtration inlet is the need for elevation differences in the storm drainage system in order to accommodate live pool storage and sand filter thickness. The minimum elevation difference needed at a site from the inflow to the outflow is 5 ft for surface sand filters and 2 to 3 ft for perimeter sand filters (Knoxville, 2003).

A minimum depth of 2 ft is required between the bottom of the sand filter and the elevation of the seasonally high water table for surface sand filters with exfiltration (i.e,. earthen structure). While there are no restrictions on the type of soils, Group "A" soils are generally required to allow exfiltration (Atlanta Regional Commission, 2001).

Sizing

Many guidelines recommend sizing the filter bed using Darcy's Law, which relates the velocity of fluids to the hydraulic head and coefficient of permeability of a medium. Hydraulic calculations based on Darcy's Law used to establish the filter area of a sand filter allow flow-through of the treatment volume within the desired time frame, typically 40 to 48 h (City of Austin, 1988; VA DCR, 1999). The State of Florida uses more complex falling-head computations and allows a drawdown time of up to 72 h (Florida DER, 1988). However, creating storage for the full WQ_v in shallow configuration systems may result in a larger filter than the hydraulic calculations would indicate (VA DCR, 1999).

The Austin Sand Filter Formula (City of Austin, 1988) derived from Darcy's Law to size sand filters is given as:

$$A_f = \frac{I_a H d_f}{k(h + d_f)t_f} \qquad \textbf{(3-15)}$$

where
A_f = surface area of sand bed (acres or ft^2);
I_a = impervious drainage area contributing runoff to the basin (acres or ft^2);
H = runoff depth to be treated (ft);
d_f = sand bed depth (ft);
k = coefficient of permeability for sand filter (ft/h);
h = average depth (ft) of water above surface of sand media between full and empty basin conditions (½ max. depth) (City of Austin, 1996); and
t_f = time required for runoff volume to pass through filter media (h)

A BMP drawdown time (t_f) of 40 h allows the filter to fully drain down and dry out to maintain an aerobic environment between storms. Typical values for k are shown in Table 3-10.

Table 3-10. Coefficient of Permeability k Values for Stormwater Filtering Practices

Filter Medium	Coefficient of Permeability (ft/d)
Sand	3.5
Peat/Sand	2.75
Compost	8.7

(CWP, 1996a)

The permeability of sand shown in Table 3-10 is extremely conservative, but is widely used since it is incorporated in the design guidelines of the City of Austin(City of Austin, 1988; City of Austin, 1996). When the sand is initially installed, the permeability is so high (over 100 ft/d) that generally only a portion of the filter area is required to infiltrate the entire volume, especially in a "full sedimentation" Austin design where the capture volume is released to the filter basin over 24 h. This methodology results in a filter bed area that is oversized when new and the entire water quality volume is filtered in less than a day with no significant height of water on top of the sand bed. The Austin design variations are still preferred where there is sufficient space, because they lack a permanent pool, which eliminates vector concerns. Consequently, the simple rule of thumb is adequate for sizing the filter area.

For filters with full sedimentation protection (sedimentation basin containing full WQ_v with 24 h drawdown to filter), $k = 3.5$ ft/d (0.146 ft/h), and $t_f = 40$ h, the sand filter formula reduces to:

$$A_{f(FS)} = \frac{310 I_a d_f}{(h + d_f)} \qquad \textbf{(3-16)}$$

where
A_f is in ft^2 and I_a is in acres.

For filters with partial sedimentation protection (sediment chamber containing 20% of WQ_V with free hydraulic flow to filter), $k = 2.0$ ft/day (0.0833 ft/h) and $t_f = 40$ h, the formula reduces to:

$$A_{f(PS)} = \frac{545 I_a d_f}{(h + d_f)} \qquad \textbf{(3-17)}$$

Where
A_f is in ft^2 and I_a is in acres.

Capture volume

The facility should be sized to capture the required water quality volume, preferably in a separate pretreatment sedimentation basin.

Geometry

The water depth in the sedimentation basin when full should be at least 2 ft and no greater than 10 ft. A fixed vertical sediment depth marker should be installed in the sedimentation basin to indicate when 20% of the basin volume has been lost because of sediment accumulation.

Basic Components

Surface sand filters generally employ the following layers, from top to bottom: sand, geotextile and an underdrain system. Runoff discharging to the sand filter must be pretreated (e.g., a presettling basin) to remove debris and other gross solids and any oil from high-use sites. The type of pretreatment device will depend on the type of pollutants present. The length-to-width ratio of the presettling basin should be 3:1 and the recommended depth varies from 3 to 6 ft.

Inlet structures such as flow spreaders, weirs or multiple orifice openings should be designed to minimize turbulence and spread the flow uniformly across the surface of the filter media.
Stone riprap or other dissipation devices should be installed to prevent gouging of the sand media and promote uniform flow. Offline outlet structures are typically sized for the 15-min peak flow of a 2-yr, 24-h storm.

An impermeable liner (clay, geomembrane or concrete) may be required under the filter to protect groundwater or where underflow could damage structures. If the impermeable liner is not required, a geotextile liner should be installed, unless the bed has been excavated to bedrock (Metro Council, 2001).

The sand filter is typically constructed with 18 in. of sand overlying 6 in. of gravel. The sand and gravel media are separated by permeable geotextile fabric and the gravel layer is situated on geotextile fabric. Four-in.perforated PVC pipe is used to drain captured flows from the gravel layer. A minimum of 2 in. of gravel must cover the top surface of the PVC pipe.

Sand Specification

The sand in a filter must consist of medium-sized sand that meets the size gradation (Table 3-11). A laboratory analysis to determine the sand's hydraulic conductivity K is also highly recommended. The designer should then adjust this number to account for conditioning of the sand during operation.

Table 3-11. Sand Medium Specification

U. S. Sieve Number	Percent Passing
4	95-100
8	70-100
16	40-90
30	25-75
50	2-25
100	<4
200	<2

(King County, 1998)

Underdrain Systems

Several types of underdrains may be used: a central collector pipe (with lateral feeder pipes or a geotextile drain strip in an 8 in. gravel backfill or drain rock bed) or a longitudinal pipe in an 8 in. gravel backfill or drain rock with a collector pipe at the outlet end.

Hydraulically, the system is typically sized for the 15 min peak flow from a 2-yr, 24-h storm, with 1 ft of head above the invert of the upstream end of the collector pipe. Yet, local sizing requirements should be used when available.

A geotextile fabric must be used between the sand layer and drain rock or gravel and placed so that 1.0 in. of drain rock or gravel is above the fabric. Drain rock should be 1.5 to 0.75 in. rock or gravel backfill, washed free of clay and organic material.

Cleanout wyes with caps or junction boxes must be provided at both ends of the collector pipes. Cleanouts must extend to the surface of the filter. A valve box must be provided for access to the cleanouts.

Impermeable Layers

Impermeable liners such as clay, concrete, or geomembrane should be used when non-conventional soluble pollutants such as metals or organics are present, and where the underflow could cause problems with structures or groundwater. Clay liners should have a minimum thickness of 12 in. and meet the specifications in Table 3-12.

A geomembrane liner should be at least 30 mils thick and ultraviolet resistant. It should be protected from puncture, tearing and abrasion by installing geotextile fabric on the top and bottom of the geomembrane.

Concrete liners may also be used for basins less than 1,000 ft^2 in area. Concrete should be 5 in. thick (Class A or better) and reinforced by steel wire mesh. The concrete should have a minimum 6 in. compacted aggregate base consisting of either coarse sand and river stone or crushed stone or its equivalent with diameter of 0.75 to 1 in., when the underlying soil is clay or has an unconfined compressive strength of 0.25 ton/ft^2 (Metro Council, 2001).

If an impermeable liner is not provided, an analysis should be made of possible adverse impacts of seepage zones on groundwater and nearby built areas. Sand filters without impermeable liners should not be built on fill sites, and should be located at least 20 ft downslope and 100 ft upslope from building foundations (Metro Council, 2001).

Table 3-12. Clay Liner Specifications

Property	**Test Method**	**Unit**	**Specification**
Permeability	ASTM D-2434	cm/sec	1×10^{-6} max
Plasticity index of clay	ASTM D-423 & D-424	percent	Not less than 15
Liquid limit of clay	ASTM D-2216	percent	Not less than 30
Clay particles passing	ASTM D-422	percent	Not less than 30
Clay compaction	ASTM D-2216	percent	95% of Standard Proctor Density

(WA DOE, 2001)

Underground Filters

Although, in general, sand filter design criteria apply to underground filters as well, additional specific recommendations for underground filters are as follows:

- One ft of sediment storage in the presettling cell should be provided.

- The retaining baffle for oil/floatables in the pre-settling cell must extend at least 1 ft above to 1 ft below the design flow water level, and be spaced a minimum of 5 ft horizontally from the inlet. Provision for the passage of flows in the event of plugging must be provided. Access opening and ladder must be provided on both sides of the baffle.

- The inlet flow distribution should be optimized with minimal sand bed disturbance. One recommendation is to provide a maximum of 8 in. of distance between the top of the spreader and the top of the sand bed. Flows may enter the sand bed by spilling over the top of the wall into a flow spreader pad. Alternatively a pipe and manifold system may be used. Any pipe and manifold system must retain the required dead storage volume in the first cell, minimize turbulence, and be readily maintainable. Multiple inlets are recommended to minimize turbulence and reduce local flow velocities.

- Erosion protection must be provided along the first foot of the sand bed adjacent to the spreader. Geotextile fabric secured on the surface of the sand bed, or an equivalent method may be used. A dewatering gate valve should be constructed just above the sand bed and removable sand panels must be provided over the entire sand bed.

- To prevent anoxic conditions, a minimum of 24 ft^2. of ventilation grate must be provided for each 250 ft^2 of sand bed surface area. For sufficient distribution of air flow across the sand bed, grates may be located in one area if the sand filter is small, but placement at each end is preferred. Small grates may also be dispersed over the entire sand bed area.

Organic Filters

The organic filter is a design variant of the surface sand filter, which uses organic materials such as leaf compost or a peat/sand mixture as the filter media. The organic material enhances pollutant removal by providing adsorption of contaminants such as soluble metals, hydrocarbons, and other organic chemicals (Atlanta Regional Commission, 2001). Additional specific recommendations for organic filters are as follows: The type of peat used is critically important. Fibric peat, in which undecomposed fibrous organic material is readily identifiable, is preferred. Hemic peat containing more decomposed material may also be used. Sapric peat, made up largely of decomposed matter, is not recommended. They are typically used on relatively small sites (up to 10 acres), to minimize potential clogging. The minimum head requirement (the elevation difference needed at a site from the inflow to the outflow), 5 to 8 ft, is higher than the surface sand filter.

Two typical media bed configurations are the peat/sand and the compost filter. Both variants utilize a gravel underdrain system. The peat filter includes an 18 in. 50/50 peat/sand mix over a 6 in. sand layer and can be optionally covered by 3 in. of topsoil and vegetation. The compost filter has an 18 in. compost layer.

Figures 3-9 to 3-12 show the schematic of a surface sand filter, perimeter sand filter, underground sand filter, and organic filter, respectively.

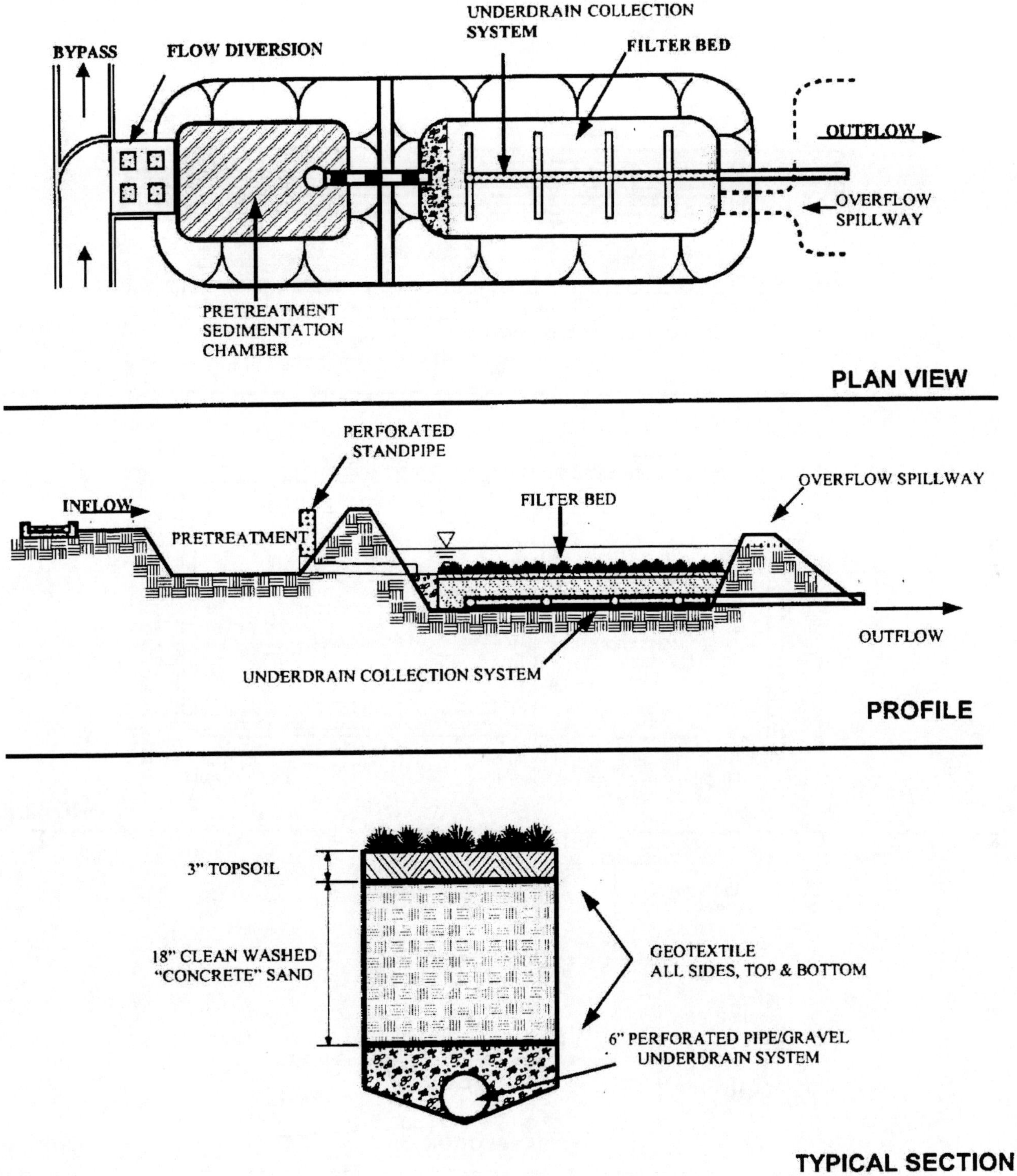

Figure 3-9. Schematic of a Surface Sand Filter (MDE, 2000)

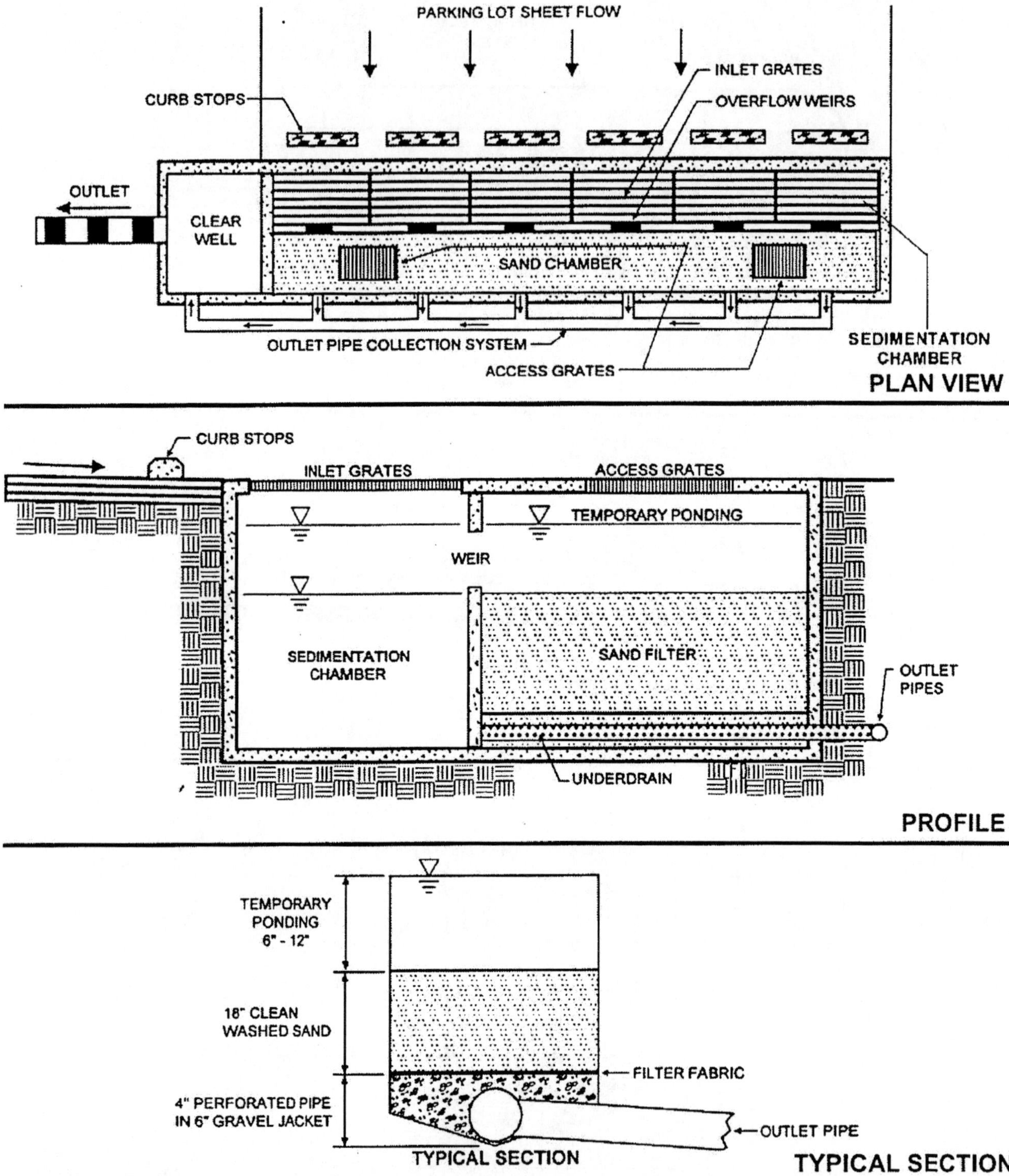

Figure 3-10. Schematic of a Perimeter Sand Filter (MDE, 2000)

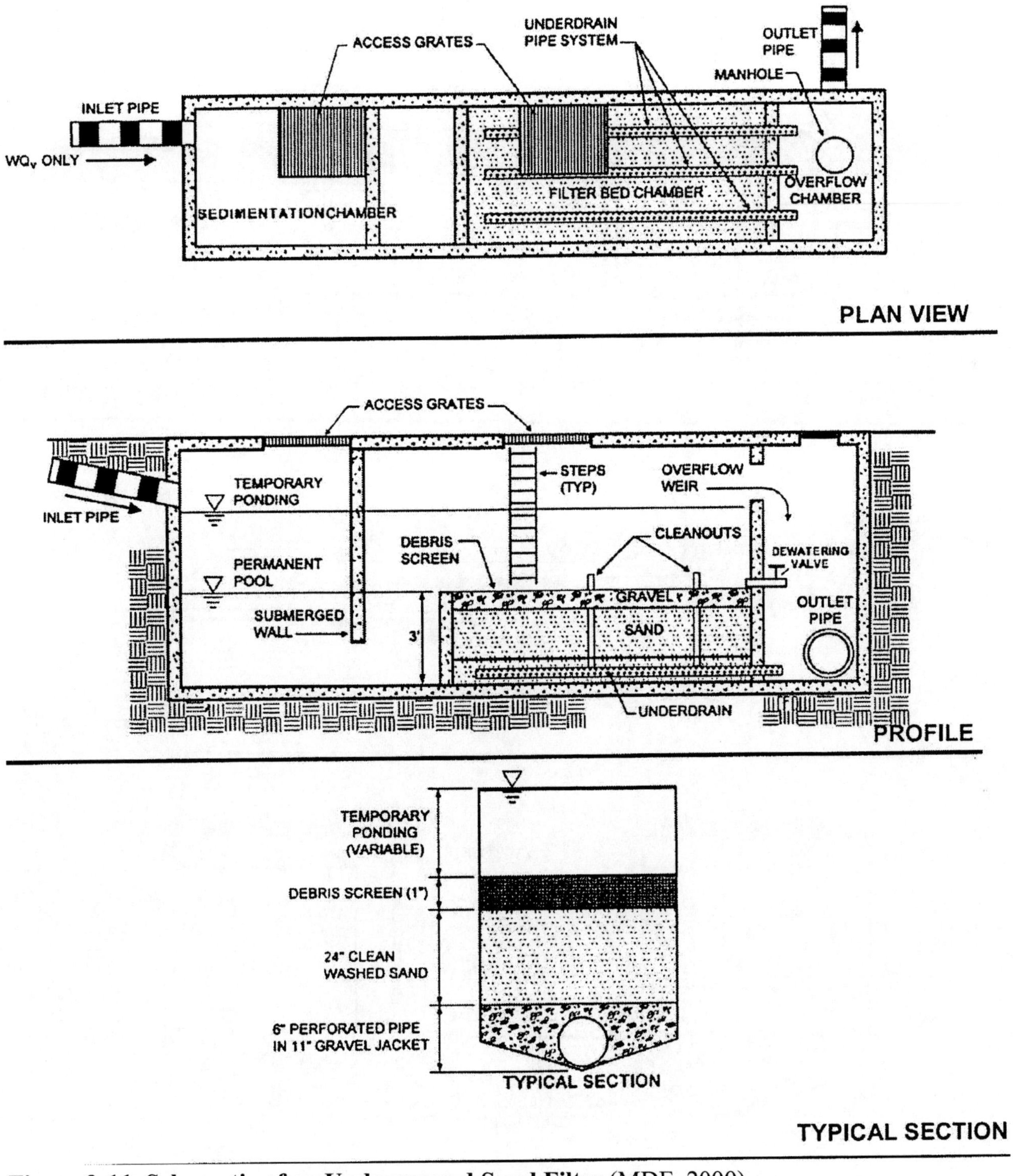

Figure 3-11. Schematic of an Underground Sand Filter (MDE, 2000)

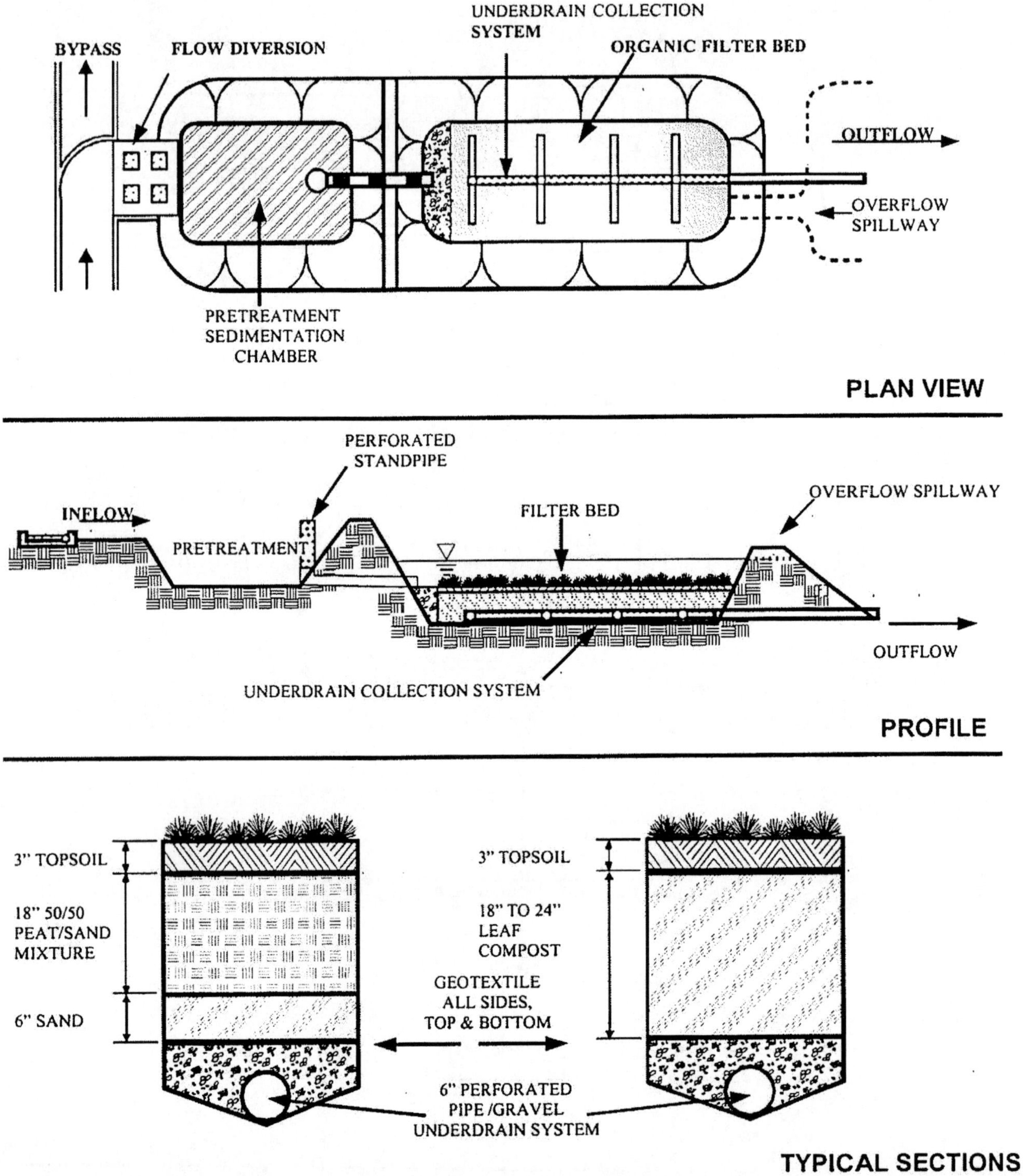

Figure 3-12. Schematic of an Organic Filter (MDE, 2000)

References

Atlanta Regional Commission. (2001). *Georgia Stormwater Management Manual Volume 2: Technical Handbook.* Atlanta Regional Commission, www.georgiastormwater.com

California Department of Transportation (Caltrans). (2002). *Proposed Final Report: BMP Retrofit Pilot Program.* CTSW-RT-01-050.

California Stormwater Quality Association (CASQA). (2003). *Stormwater Best Management Practice Handbook: New Development and Redevelopment.* www.cabmphandbooks.com.

Chow, V.T. (1959). *Open-Channel Hydraulics.* McGraw Hill, New York, NY.

(City of Austin). (1988). *Environmental Criteria Manual.* Austin, TX.

(City of Austin). (1991). *Water Quality Management. Environmental Criteria Manual.* Austin, TX.

(City of Austin). (1996). *Design of Water Quality Controls.* Austin, TX.

Clar, M., B.J. Barfield, and S. Yu. (2003). *Considerations in the design of treatment best management practices (BMPs) to improve water quality.* EPA 600/R-03/103. U.S. Environmental Protection Agency, Office of Research and Development, Cincinnati, OH. www.epa.gov/ORD/NRMRL/Pubs/600r03103/600r03103.htm October 2003.

Claytor, R.A. and T.R. Schueler. (1996). *Design of Stormwater Filtering Systems.* Center for Watershed Protection and Chesapeake Research Consortium, Ellicott City and Solomons, MD.

Center for Watershed Protection (CWP). (1996a). *Design of Stormwater Filtering Systems.* Ellicott City, MD.

Center for Watershed Protection (CWP). (1996b). *Ditches or Biological Filters? Classifying the pollutant removal performance of open channels. In Watershed Protection Techniques, Volume 2 No. 2.* Ellicott City, MD.

Florida DER. (1988). *The Florida Development Manual: A Guide to Sound Land and Water Management.* Florida Department of Environmental Regulation, Tallahassee, FL.

Guo, C.Y. and B.R. Urbonas. (1995). *Special Report to the Urban Drainage and Flood Control District on Stormwater BMP Capture Volume Probabilities in United States,.* Denver, CO.

Horner, R.R. (1988). *Biofiltration Systems for Storm Runoff Water Quality Control.* Washington State Department of Ecology, Seattle, WA.

King County Department of Natural Resources (King County). (1998). *King County Washington Surface Water Design Manual.* Seattle, WA.

City of Knoxville (Knoxville). (2003). *Best Management Practices (BMP) Manual.* Knoxville, TN.

Maryland Department of the Environment (MDE). (2000). *Maryland Stormwater Design Manual Volumes I & II.* Baltimore, MD. http://www.mde.state.md.us.

(Metro Council). (2001). *Minnesota Urban Small Sites BMP Manual. Stormwater Best Management Practices for Cold Climates.* St. Paul, MN.

Metropolitan Washington Council of Governments (MWCOG). (1979). *Controlling Stormwater Runoff in Developing Areas: Selected Best Management Practices.* Washington, DC.

Northern Virginia Planning District Commission (NVPDC). (1979). *Guidebook for Screening Urban Nonpoint Pollution Management Strategies.* Annandale, VA.

Northern Virginia Planning District Commission (NVPDC). (1987). *BMP Handbook for the Occoquan Basin (Draft).* Annandale, VA.

Northern Virginia Planning District Commission and Engineers and Surveyors Institute (NVPDC and ESI). (1992). *Northern Virginia BMP Handbook: A guide to planning and designing best management practices in northern Virginia*. Annandale, VA.

Schueler, T.R. (1987). *Controlling Urban Runoff: A practical manual for planning and designing urban BMPs*. Metropolitan Washington Council of Governments, Washington, DC.

Schueler, T.R. (1992a). *Design of Stormwater Wetland Systems: guidelines for creating diverse and effective stormwater wetlands in the mid-Atlantic Region*. Metropolitan Washington Council of Governments, Washington, DC.

Schueler, T.R. (1992b). *Design of Stormwater Pond Systems*. Metropolitan Washington Council of Governments,, Washington, DC.

Schueler, T.R., P.A. Kumble, and M.A. Hearty. (1992). *A Current Assessment of Urban Best Management Practices: Techniques for reducing nonpoint source pollution in the coastal zone*. Metropolitan Washington Council of Governments, Washington, DC.

Strecker, W.E., J.M. Kersnar, and E.D. Driscoll. (1990). *The use of wetlands for controlling stormwater pollution*. Woodward-Clyde Consultants. Prepared for the United States Environmental Agency.

United States Environmental Protection Agency (U.S. EPA). (1977). *Process Design Manual for Land Treatment of Municipal Wastewater. EPA-625-1-77-008*. Washington, DC.

United States Environmental Protection Agency (U.S. EPA). (1983). *Results of the Nationwide Urban Runoff Program. Volume I - Final Report. EPA-832-R-83-112*. Washington, DC.

United States Environmental Protection Agency (U.S. EPA). (1996). *Overview of the Stormwater Program. EPA-833-R-96-008*. Washington, DC.

Urban Drainage and Flood Control District (UDFCD). (2002). *Urban Storm Drainage Criteria Manual: volume 3 - best management practices*. Denver, CO. www.udfcd.org.

Urbonas, B.R. and P. Stahre. (1993). *Stormwater - Best Management Practices Including Detention*. Prentice Hall, Englewood Cliffs, NJ.

Commonwealth of Virginia Department of Conservation and Recreation (VA DCR). (1999). *Virginia Stormwater Management Handbook, Volume I,*. Richmond, VA.

Washington State Department of Ecology (WA DOE). (1991). *Stormwater Management Manual for the Puget Sound Basin - Public Review Draft*. Olympia, WA.

Washington State Department of Ecology (WA DOE). (2001). *Stormwater Management Manual for Western Washington. Volume V - Runoff Treatment BMPs*. http://www.ecy.wa.gov/programs/wq/stormwater/index.html.

Water Environment Federation and American Society of Civil Engineers (WEF and ASCE). (1998). *Urban runoff quality management*. 1-57278-039-8 and 0-7844-0174-8. Alexandria, VA.

Wyoming Department of Environmental Quality (WY DEQ). (1999). *Urban Best Management Practices for Nonpoint source Pollution*.

Yousef, Y., M. Wanielista, and H. Harper. (1985). *Best Management Practices: Removal of highway contaminants by roadside swales*. Prepared for Florida Department of Highway Transportation, Tallahassee, FL.

4 BMP Monitoring

Bethany Madge

CONTENTS

4.0 Introduction

Implementation of an effective BMP monitoring program is not a straight-forward task. BMPs by definition are devices, practices, or methods used to manage stormwater runoff. This umbrella term lumps widely varying techniques into a single category. Some BMPs, especially nonstructural BMPs are pollution prevention techniques, such as low-impact development techniques (LIDs), or source control techniques, such as street sweeping. These BMPs cannot be directly associated with an influent and effluent that isolates their effects. Other BMPs are strictly governed by diffuse or nonpoint source concepts, such as buffer strips, while larger regional BMPs such as downstream detention ponds, behave more like intermittent point sources than traditional diffuse or nonpoint sources. Thus, a wide variety of underlying conditions may exist, making a one-size-fits-all approach to BMP monitoring infeasible.

Great variability in stormwater properties and the associated runoff complicates BMP monitoring further. Precipitation varies in time and space. Stormwater pollutants can be carried into the receiving water system by the precipitation itself (wet deposition) and/or picked up as it flows across surfaces with or without conveyance by man-made or natural drainage channels. Air quality, land use, drainage systems, and geology characteristics are all nonuniform, again leading to temporal and spacial variation concerning stormwater pollutant loads. An effective BMP monitoring program must incorporate this variability to produce reliable data. Given these difficulties, it is no surprise that a report produced by the Governmental Advisory Committee in 2000 on current monitoring practices in the U.S. noted that data gaps are prevalent and "particularly serious for nonpoint sources" (U.S. EPA, 2002a). Nonpoint sources are recognized as the major contributor to the pollution of the nation's waters, particularly since regulatory measures made vast reductions in point source pollution. BMPs are the primary tools used to mitigate the deleterious effects of nonpoint sources on receiving waters, yet there is little evidence that BMPs are meeting their projected goals (Strecker and Urbonas, 2001). Therefore, high quality BMP monitoring programs are an important piece in completing the current picture of the nation's water quality and making steps towards improvements.

There are many types of BMP monitoring: implementation, trend, effectiveness, pre-BMP stormwater monitoring to establish BMP design, project validation, whole system effectiveness, and compliance monitoring. The goals and objectives of a given monitoring program will dictate which type of monitoring is appropriate. The focus of this section is effectiveness monitoring. However, pre-BMP stormwater monitoring to establish BMP design may be the most practiced type of BMP monitoring, therefore, important issues specific to this type of monitoring will be identified where applicable. BMP implementation and trend monitoring are covered in detail in the U.S. EPA report *Techniques for Tracking, Evaluating, and Reporting the Implementation of Nonpoint Source Control Measures - Urban* (U.S. EPA, 2001). Although the different types of monitoring have their own distinct set of characteristics, there is a great deal of overlap between them. Accordingly, many issues discussed in this section are not exclusive to effectiveness monitoring.

4.1 CURRENT MONITORING PRACTICES

There are four main monitoring approaches employed to assess BMP effectiveness (Strecker *et al.*, 2000). The most popular approach is input/output sampling that is used with new, existing, or retrofitted structural BMPs. A second BMP monitoring approach is before/after sampling. This approach can be used in new or retrofit BMP situations, but is most often used with nonstructural or other BMPs that lack an inflow/outflow. Upstream/downstream monitoring can be used to assess the impact of a single BMP's effluent or an untreated stormwater input on its receiving stream (FHWA, 2000). The final BMP monitoring approach is control watershed comparison. Although sometimes useful for evaluating nonstructural BMPs where before data was not collected or structural BMPs without defined inlets (e.g., vegetative filter strips), the control watershed comparison approach is rarely used due to the difficulty in finding a watershed with similar contributing factors to serve as the control.

Thorough water and pollution loading budgets are important to a robust BMP monitoring program. Stormwater flow measurements are of critical importance, yet they are often the cause of error in efficiency calculations (U.S. EPA, 1999; FHWA, 2001). Dry-weather flows, groundwater, and direct precipitation can also contribute to both hydraulic and pollutant loading, although they are rarely considered (Reinelt and Horner, 1995; Rushton, 1998; GeoSyntec and ASCE, 2002). Long-term studies that include seasonal effects on flow and pollutant patterns are rare as well (Reinelt and Horner, 1995), even though rainfall/flow and pollutant build-up vary over time. Other problems often associated with poor quality flow data include a lack of equipment maintenance and calibration, and the neglect of bypass flows (Clary *et al.*, 2001; GeoSyntec and ASCE, 2002). These factors can lead to either an over or underestimation of actual BMP efficiencies.

Collecting representative samples is another important component of a robust BMP monitoring program. There are many techniques for stormwater sample collection. The simplest is obtaining a grab sample. Composite samples include time-proportional sampling and three types of flow-proportional sampling: (1) constant time - volume proportional to flow; (2) constant time - volume proportional to flow-volume increment; and (3) constant volume - time proportional to flow-volume increment (GeoSyntec and ASCE, 2002). BMPs that do not have a clearly defined inflow and outflow, such as biofiltration, greenroofs, and vegetated filter strips, present unique difficulties in collecting representative samples. Flows into and out of these systems must be concentrated into a single or otherwise representative outfall before collection (U.S. EPA, 1999).

In addition to being representative in how it was collected, the sample must also be representative in what parameters are measured. Most BMP monitoring studies sample for selected chemical and/or physical species such as total phosphorus and total suspended solids (TSS). However, these parameters do not always provide a representative picture of BMP effectiveness. Although TSS has proven useful in wastewater monitoring, it has been found "fundamentally unreliable" for evaluating natural waters (Rushton, 2002). Biochemical oxygen demand is another traditional wastewater parameter often measured in BMP monitoring programs that has been criticized as unrepresentative for stormwater (Rushton, 2002). A key parameter for evaluating BMP effectiveness that is usually neglected is gross or coarse pollutants such as floatables or litter debris (e.g., beverage containers,

Styrofoam) and organic debris (e.g., leaves, grass clippings, and twigs) (England and Rushton, 2003). Biological and physical indicators of the ecological health of the receiving water may provide more realistic information on BMP effectiveness, depending on the objectives of the BMP monitoring program, yet these parameters are not historically measured (U.S. EPA, 1999).

Once data is collected during a BMP monitoring program, it must be analyzed. GeoSyntec and ASCE (2002) list ten different methods that have been applied to BMP monitoring data to assess effectiveness. Every method employs its own set of assumptions and as a result produces its own level of efficiency. This alone is a significant contributor to the wide range of reported BMP effectiveness. For example, an exercise that applied three of the most common data analysis methods to the same set of data resulted in percent removals from 48 to 66% (Strecker *et al.*, 2000). The most commonly used method of data analysis is event mean concentration (EMC) (U.S. EPA, 1999). An EMC is a statistical parameter representing the average concentration of a constituent over the course of a single storm event. It can be calculated by dividing the constituent's flow-proportional average mass by the total runoff volume. Alternatively, samples can be collected on a flow-proportional basis to build a flow-weighted composite sample for analysis. Concentrations of such a composite sample are directly assumed to be EMCs. BMP efficiency is often then determined based on percent removals between the influent and effluent EMCs on a storm-by-storm basis. This storm-by-storm pairing of influent and effluent may be problematic for BMPs with a permanent volume, such as retention (wet) ponds because the simultaneous inflow and outflow may not be related to the same storm event (Clary *et al.*, 2001). It is also problematic as it weighs all storms equally. One final reason this analysis may be inappropriate is that percent removals do not tell the whole picture and therefore can be misleading. For instance, one study of a detention pond in Florida found that the removal efficiency actually decreased after a modification to increase the residence time, and thus one could conclude that the modification resulted in worse performance (Strecker *et al.*, 2000). However, a closer look at the data revealed that the effluent EMCs after the modification were less than before the modification, but for unknown reasons the influent EMCs were also lower. Thus, the comparative differences between the influent and effluent were greater before the modification than after, leading to the erroneous conclusion. As shown by this example, emphasis on percent removal is unjustified and it cannot stand alone as a sole measure of BMP effectiveness (GeoSyntec and ASCE, 2002).

Typically, data analysis of BMP monitoring programs also fail to consider statistical validation (Strecker *et al.*, 2000). This point has been regarded by some as being "the most frequently overlooked factor" in a BMP monitoring program (GeoSyntec and ASCE, 2002). Obtaining statistically valid results begins with good planning to assure that the number of storms sampled is sufficient to draw valid conclusions at the specified level of confidence. As with any field experiments, unusual activities within the watershed or uncharacteristic rainfall events cannot be controlled and if there are too few sampling events, these incongruities can bias effectiveness results (Taylor and Wong, 2002b). Thus, the number of sampling events necessary to obtain statistical confidence may become cost prohibitive. After data analysis is completed, the statistical significance of a BMP monitoring program's conclusions should always be reported.

A final downfall of BMP monitoring programs is that details that contribute to the observed effectiveness of a BMP are often overlooked (Strecker *et al.*, 2000). Watershed characteristics that may affect BMP performance include watershed area, percent imperviousness, land-use breakdown, soil types, and rainfall characteristics. BMP design characteristics and maintenance activities also effect BMP performance. However, it is wise not to rely solely on pre-construction/implementation design plans, as "BMPs will not necessarily be built and/or maintained as designed and approved" (Taylor and Wong, 2002b). All of these variations result in a nearly infinite number of combinations that further complicate the evaluation and comparison of BMP monitoring results (U.S. EPA, 1999).

As the above discussion illustrates, most BMP monitoring studies reported lack the rigor required to compare and pool individual studies. Pooling of individual studies is necessary to make any definitive assessments of BMP effectiveness. Rigorous BMP monitoring programs become complex quickly. Consequently many BMP monitoring programs produce insufficient or unsound data, in part due to poor experimental design (Marsalek and Kok, 2000). Thus, extrapolation of BMP effectiveness results to unmonitored BMPs and areas under consideration for BMP construction/implementation is impossible or unreliable at best. In these situations it is always best to perform baseline monitoring of the stormwater runoff characteristics and tailor the BMP design to the specifics of each individual location.

In the past, the field of BMP effectiveness monitoring has suffered from a lack of extensive experience and good quality guidance. More recently, GeoSyntec and ASCE (2002) produced a reasonably comprehensive guidance manual for the monitoring of urban stormwater BMP performance. The main goal of this manual was to assist researchers in producing robust and consistent BMP monitoring results to expand the National Stormwater BMP Database (accessible online at http://www.bmpdatabase.org/). In addition to suggesting preferred methods of sampling, measuring flows, and analyzing data, the manual includes many data collection tables covering everything from watershed characteristics to significant BMP design parameters individualized for the most common BMPs.

4.2 Parameter Selection for Structural BMP Monitoring Programs

There is a wide variety of parameters that could be included in a BMP monitoring program. When developing a monitoring program, the number of possible parameters can be overwhelming. In this section, potential parameters are introduced under five major parameter categories. Once familiar with the parameter options, a list of key considerations used to narrow in on an appropriate set of monitoring parameters is presented.

4.2.1 Major Parameter Categories

Five major categories of BMP monitoring parameters have been identified: (1) chemical parameters; (2) physical parameters; (3) biological parameters; (4) hydrological parameters; and (5) additional contributing factors. Traditional BMP monitoring programs have focused mainly on water quality and physical parameters (U.S. EPA, 2002a). However, a robust monitoring program will incorporate some measures from most or all five of the major categories.

4.2.1.1 Chemical Parameters

In the early 1980s, the U.S. EPA's NURP established a set of nine water quality parameters (and one physical parameter, TSS, discussed in Section 4.2.1.2) thought to fully characterize urban stormwater (GeoSyntec and ASCE, 2002). These parameters, listed below, are still the most widely applied parameters in BMP monitoring programs.

- Biochemical Oxygen Demand (BOD)
- Chemical Oxygen Demand (COD)
- Copper (Cu)
- Lead (Pb)
- Zinc (Zn)
- Total Kjeldahl Nitrogen (TKN)
- Nitrate + Nitrite Nitrogen ($NO_2^- + NO_3^-$)
- Total Phosphorus (TP)
- Soluble (or ortho-) Phosphorus (SP)

Since the 1980s, the BMP scientific community has re-evaluated this list. In 1994, it was suggested by Strecker that ortho-phosphate was a more useful parameter than soluble phosphorus because ortho-phosphate approximated the bioavailable fraction phosphorus (Strecker, 1994). As discussed above, BOD has been criticized as being an inappropriate measure for stormwater. Individual nitrogen species recycle rapidly, thus their measurement may not be beneficial (Strecker, 1994; Rushton, 2002). Measurement of total metals may not provide a representative indication of toxicity because some metals are only toxic in their soluble or unbound states. Also, hardness is often used in metals standards for toxicity, and without this measurement the true consequence of metals concentrations cannot be fully assessed (Rushton, 2002). After reviewing completed BMP monitoring studies for cost-effectiveness and the generation of meaningful results, Strecker updated NURP's list of recommended water quality parameters (Strecker *et al.*, 2000). The updated list reflected the issues discussed above by excluding BOD and TKN, and adding total organic carbon (TOC), total hardness, ammonia-nitrogen (NH_3-N), and Cadmium (Cd). Both the total and soluble forms of each metal (Cd, Cu, Pb, and Zn) were recommended for analysis. Additional chemical parameters that are measured less often but can be relevant in certain situations are oil and grease, polycyclic aromatic hydrocarbons (PAHs), potentially toxic pesticides and herbicides, and other organic chemicals (Strecker, 1994; Pitt *et al.*, 1995; NJDA *et al.*, 2000; GeoSyntec and ASCE, 2002; Rushton, 2002).

In baseline monitoring, analyzing for a wide variety of chemical parameters from nutrients to organics to metals will help identify target parameters for BMP design. It is also important, particularly for baseline monitoring, to evaluate particulate-associated versus dissolved fractions of chemical pollutants, as their modes of removal and transformation within a BMP are often significantly different.

Monitoring chemical parameters provides specific information about BMP water quality. Potential effects on receiving waters can be predicted by chemical concentrations and their quantitative nature provides for a clear method of evaluating BMP performance when inflow and outflow are compared. However, direct measurement of pollutants is time-specific. Chemical data is highly variable and thus "100% valid only at the precise moment the sample was taken" (NJDA *et al.*, 2000). Water quality trends and cumulative effects of chronic, low concentrations are not readily observed when monitoring chemical parameters alone.

4.2.1.2 Physical Parameters

The physical characteristics of runoff are key to BMP monitoring programs. Not only are the physical parameters themselves, especially TSS, often the main specified management goal of BMPs, they also act as an indicator for, and can become carriers of, many of the chemical pollutants discussed in Section 4.2.1.1 (Ruby and Kayhanian, 2003). As such, TSS is often the only parameter measured in BMP monitoring programs because it is believed to indicate overall water quality (Rushton, 2002). However, caution must be used when making this assumption because the correlation between TSS and other parameters is most often not strong enough to eliminate the need to specifically address other significant parameters (Strecker, 1994). Furthermore, TSS may not be a representative measure of the physical nature of runoff because larger sediments settle out quickly and thus the procedure for TSS measurement will not include this portion of the sediment mass. In a study supported by the U.S. Geological Survey (USGS), it was found that TSS measurements may underestimate the actual sediment load by an order of magnitude or more, particularly in the influent (GeoSyntec and ASCE, 2002). Thus, a new measurement of sediments, the suspended-sediment concentration (SSC) was developed and is now part of the USGS policy on water monitoring. SSC differs from TSS solely by the specification that SSC must use the entire sample, while TSS measurement may (and in most situations does) use a sub-sample of the originally collected sample. Although SSC is considered by most to be a more appropriate measure for BMP monitoring programs, TSS measurement generally is recommended due to its historical use to enable comparison with existing data and development of trends (GeoSyntec and ASCE, 2002). The freedom to use sub-samples in TSS measurement also lends itself more easily to the analysis of composite samples often used in stormwater monitoring. Further, using SSC to determine BMP performance (by percent removal) with respect to solids can bias the results, while TSS will provide a much more conservative measure of BMP performance. This bias is less important if other measures, like average effluent concentrations, are used to determine BMP effectiveness.

Gross solids, such as litter, trash, and other debris, are another significant physical parameter in stormwater. Although seldom measured, control of gross pollutants is principal to maintaining the aesthetic value of receiving waters. Since aesthetic quality is the one effect of untreated stormwater that is easily identifiable and understandable by the public, the effectiveness of BMPs in controlling gross pollutants should be included in a robust BMP monitoring plan. In addition to their direct impact on aesthetics, gross solids may also "degrade aquatic habitat, ...smother productive sediments, leach harmful pollutants, and cause unpleasant odors" (England and Rushton, 2003). Measurement of gross solids is difficult because they cannot be collected by the automatic samplers commonly used in stormwater monitoring (Rushton, 2002). England and Rushton (2003) have developed a method of monitoring gross pollutants as part of a BMP maintenance program, measuring volume and dry mass of gross solids during BMP clean out.

Other physical parameters that may be important to monitor include turbidity, particle size distribution, settling velocity distribution, and accumulated sediments. Particle size distribution data can be useful to estimating the BMP capture efficiency of suspended solids, particularly if particle specific gravity is measured (GeoSyntec and ASCE, 2002). Measurement of settling velocity directly is also an option which is recommended (Field and O'Connor, 1996; GeoSyntec and ASCE, 2002). Particle size distribution and specific gravity or settling velocity are important for baseline monitoring to establish BMP design and for influent samples where settleability may be a significant factor in performance variations between BMPs. Monitoring accumulated sediments may be important in BMPs that encourage particle settling. Typically, the sediments represent a sink for both physical pollutants (sediments) and their associated water quality pollutants. However, under some conditions, accumulated sediments can become a pollutant source (Kaiserli *et al.*, 2002). In areas with high heavy metals loads, the accumulated sediments may actually become toxic and subject to hazardous waste regulations (WEF and ASCE, 1998). Accumulated sediments can be measured by coring, grab sampling, or using sedimentation traps that have the added benefit of allowing estimates of sedimentation rates (Pettersson, 2001).

One final item that can be characterized as a physical parameter is physical changes to the receiving water channel caused by stormwater (WEF and ASCE, 1998). Such changes include stream down-cutting and are primarily the result of changes in the stream hydrology that may be altered by the BMP. U.S. EPA's Rapid Bioassessment Protocols include a physical habitat assessment component that may be used to monitor physical changes to the receiving water channel over time (U.S. EPA, 1999).

4.2.1.3 Biological Parameters

Biological parameters can be divided into two main groups, those that BMPs strive to prevent (i.e., direct measurement of organisms associated with pollution), and those that BMPs strive to protect (ecological effects on receiving waters). The former is still relatively rare in BMP monitoring programs (Rushton, 2002). Pathogenic microorganisms present in stormwater can lead to fish kills in extreme cases and more often beach closings. The impact of BMPs on pathogenic microorganism such as bacteria, viruses, *Cryptosporidium* and *Giardia*, are not well characterized at present due to limited monitoring data. Indicator organisms such as fecal coliform, which are traditionally used in the wastewater industry, may not be representative for stormwater applications (O'Shea and Field, 1992; Rushton, 2002). However, total and fecal coliform remain the basis of most permits and state water quality standards, and are the reasoning behind most beach closings. Thus, measurement of these indicator organisms provides practical information.

The overall goal of any BMP implementation is ultimately to preserve or restore the health of the nations's waterways. Therefore, monitoring ecological effects downstream of a BMP is considered by some a better indicator of BMP effectiveness than water quality parameters alone (Clary *et al.*, 2001). This is especially true when monitoring for a long-term response since ecological effects generally reflect trends in water quality, barring acute effects of obvious catastrophic events (NJDA *et al.*, 2000). Ecological effects on receiving waters can be assessed in two ways, toxicity testing of the BMP effluent and in-stream indices. Both provide information for assessing the combined impact of multiple stressors (U.S. EPA, 1996).

Whole water toxicity testing for BMPs removes a sample of a BMP effluent stream, places many organisms of a single species directly into the sample and monitors the mortality rates, growth rates, and other changes in behavior or overall health. The intermittent nature and variability in flow rates and concentrations unique to wet-weather flows requires special protocols for a true assessment of toxicity potential. Herricks (1996) has developed toxicity protocols for wet-weather flows that take into consideration the timescale issues associated with pulsed inputs of stormwater runoff. The Microtox® toxicity-screening procedure has also been used to evaluate stormwater (Pitt *et al.*, 1995). Toxicity testing has the advantage over in-stream indices of being performed in a controlled laboratory environment using indicator organisms whose growth characteristics are well known, such as *Daphnia pulex* or *Pimephales promelas* (NJDA *et al.*, 2000). The test usually will show negative effects, if present, within 24 to 72 hours. When combined with water quality parameter concentrations, toxicity testing can provide an indication of pollutant bioavailablity (Marsalek and Kok, 2000). However, the tests can be costly and their results may be highly variable (Marsalek and Kok, 2000; NJDA *et al.*, 2000).

In-stream indices can be developed from analysis of fish, benthic macroinvertebrates, and plant communities (U.S. EPA, 2002a). Fish are useful long-term indicators because their longer life span (3 to 4 yrs) allow for bioaccumulation of toxins in the organism's tissue (NJDA *et al.*, 2000). Fish communities can be monitored for individual species abundance, indication of deformities or illness, such as discoloration, and pathologic parameters, like presence of tumors or bioaccumulated toxin concentrations. Fish of the highest trophic level are particularly useful because "they tend to integrate changes in lower trophic levels, ...[thus reflecting] overall ecosystem condition" (U.S. EPA, 2002a). Including fish in a BMP monitoring program also can be a good way to muster public interest and support. Unlike other biological indices, fish are an organism the public can easily identify as beneficial to the watershed, especially in areas where sportfishing is popular. However, in some cases the receiving water may be too small to support a fish community. Benthic macroinvertebrates, such as snails and insect larvae, have limited mobility and shorter life spans than fish. Thus, they provide better site specific information which may enable source location and their reaction times are faster (NJDA *et al.*, 2000). Identification of macroinvertebrates is more difficult and time consuming. In addition there are seasonal trends in species types that can vary from year to year. These issues limit its applicability. Methods for assessing fish and benthic communities, detailed in U.S. EPA's Rapid Bioassessment Protocols, may be suitable for assessing BMP effectiveness (U.S. EPA, 1999). Finally, plants, being at the bottom of the food chain, can provide a direct link to pollutant levels. An example of a plant indicator is the presence/amount of algae attached to rocks (U.S. EPA, 2002a).

All three in-stream indicators (fish, benthic macroinvertebrates, and plant communities) share several disadvantages. First, in-stream indices are not specific to the effects of BMPs. Many other urban stressors contribute to a stream's biological condition, including "disruption of physical habitat, alteration of hydrologic patterns, introduction of non-indigenous biota, and widespread alteration of the landscape" (U.S. EPA, 2002a). Depending on the size of the stream as compared to the volume of BMP effluent, simple dilution effects may obscure any BMP effects. Additional factors, such as fish harvesting/stocking, rainfall or drought patterns, interfering upstream conditions, or possible groundwater interferences, can overshadow the effects of any single BMP (Strecker and Urbonas, 2001). Second, the lag between introduction (or removal) of the stressor and observed biological response is long enough that in-stream indices are inappropriate for evaluation on a storm-by-storm basis (U.S. EPA, 1996). Third, in-stream parameters exhibit a large degree of seasonality. Lastly, results of in-stream biological parameters vary greatly depending upon the characteristics of the sampling location. Streams may be both horizontally and vertically stratified due to primary and secondary currents. It is particularly difficult to obtain a representative sample of chemically reactive or particulate associated pollutants whose concentration will vary with distance from the BMP outfall. The pollutant's reaction rates (chemical and settling) in relation to the stream velocity must be taken into account when choosing the optimum sampling location for reactive pollutants. Other localized factors that may affect in-stream pollutant concentrations include riffle vs. pool and the percentage of area receiving full sunlight (U.S. EPA, 1996). Resuspension of pollutants associated with previously settled sediments is another confounding factor. For these reasons, sampling location characteristics should be well documented.

4.2.1.4 Hydrologic and Hydraulic Parameters

The foundation of a good BMP monitoring program, both baseline and effectiveness, is an accurate and representative measurement of precipitation and stormwater flow data (U.S. EPA, 1999). The hydraulic alteration of stormwater associated with human activity such as increases in peak runoff rates and total runoff volumes is a problem in and of itself, causing downstream flooding, lower base flows, increased erosion, and habitat destruction (NJDA *et al.*, 2000). The role of BMPs in mitigating these effects are a substantial and historic component of their intended function.

Defining storm and runoff hydrology is not a trivial matter. A number of hydrologic and hydraulic parameters such as antecedent conditions, pattern of precipitation intensities, precipitation durations, and total precipitation volumes, runoff rates, durations, and total volumes into and out of the BMP, all contribute to "make each storm a unique event"(Church *et al.*, 1999). Additional parameters such as possible bypass or overflow volumes and influences of soil infiltration, groundwater, dry-weather flows, and evaporation add another layer of complexity. All this variability translates into a great deal of inconsistency in reported BMP effectiveness. Therefore, measurement of hydrologic and hydraulic parameters is key to interpreting BMP monitoring data and predicting effectiveness.

The necessity of some hydrologic and hydraulic parameters is dependent on the BMP type. For BMPs designed to encourage infiltration such as infiltration basins, flow reductions are the primary mechanism of stressor alleviation. A reasonably accurate water balance including all inflows and outflows and antecedent conditions and precipitation intensities, both of which will affect infiltration capacity, are essential for these types of BMPs. Accurate flow measurements are especially important since they are often used as a basis for collecting water quality samples and required for the calculation of pollutant loadings. Precipitation volumes and durations are often significant in defining storm events appropriate for monitoring, although it is optimum to monitor all wet-weather events continuously.

The USGS report, *Basic Requirements for Collecting, Documenting, and Reporting Precipitation and Stormwater-Flow Measurements* prepared by Church *et al.* (1999), is an excellent source of guidance for collecting this type of data. Information on the actual collection of hydrologic and hydraulic parameters is discussed further in Section 4.5.2.3.

4.2.1.5 Contributing Factors

In addition to chemical, physical, biological, and hydrologic/hydraulic parameters, many other contributing factors determine the overall performance of a BMP. Watershed characteristics such as size, geographical location, land use, percent imperviousness, topography, soil characteristics, and upstream nonstructural BMPs should be collected during a comprehensive BMP monitoring program (U.S. EPA, 1996; GeoSyntec and ASCE, 2002). BMP design characteristics are also important contributing factors. Variables such as length, width, depth, storage volume, inlet and outlet design, vegetation, and bottom lining or soil infiltration characteristics should be documented. The complete set of BMP design characteristics is specific to the type of BMP being monitored. GeoSyntec and ASCE (2002), in their *Urban Stormwater BMP Performance Monitoring: A guidance manual for meeting the national stormwater BMP database requirements* have included detailed tables and photocopy-ready forms of BMP characteristics useful for evaluating effectiveness for many of the more common BMPs. Additional BMP characteristics include age and maintenance practices.

Contributing environmental characteristics, such as temperature, dissolved oxygen (DO), solar intensity, and pH, are also useful in the analysis of monitoring data. Temperature can affect process reaction rates within the BMP. Also, heat is specifically mentioned as a pollutant in the Clean Water Act's definition of pollutant (Parikh, 2003). Thus, increased temperatures, produced by areas collecting runoff from large paved areas or by the quiescent water of retention-type BMPs during dry periods, could be a pollutant (NJDA *et al.*, 2000). DO concentrations may be an important parameter to monitor in detention-type systems. Pollutants such as phosphorus and some metals (Cd, Cu, Pb, and Zn) that settle to the bottom with suspended solids may leach into the water column during periods of low DO concentrations (Rushton, 1998). As with temperature, low DO can also be considered a pollutant. The oxygen dependent organisms within a receiving water, such as fish, can be severely impaired or killed by prolonged low DO conditions. Hardness, as discussed earlier in Section 4.2.1.1, is another contributing environmental characteristic that is particularly useful for metals analysis.

4.2.2 Key Considerations for Selecting Appropriate Parameters

The number of possible parameters that may be measured in a BMP monitoring program is extensive. It is often impractical to measure all the parameters discussed in Section 4.2.1, and there is no "one-size-fits-all" set of parameters that will satisfy the objectives of every monitoring program. The planning phase of a BMP monitoring program must include the selection of appropriate parameters. The following questions are a list of key considerations that may be useful during the parameter selection process.

What parameters are required to meet the monitoring program objectives and goals?
If the monitoring program objectives are well defined, this may be the only question that needs asking. The objectives and goals will depend in part on the given type of monitoring program. Parameters that are appropriate to meet the objectives and goals of a baseline monitoring program to establish BMP design may be different from those of an effectiveness monitoring program. Further, BMPs may be implemented for many reasons, which could include regulatory compliance, total maximum daily load (TMDL) compliance, protection of sensitive ecosystems, etc. These reasons typically define the monitoring objectives and goals and, in turn, the list of appropriate parameters (Caltrans, 2000). For example, if a BMP is constructed to aid in the compliance with a TMDL for phosphorus, phosphorus must be measured.

What resources are available for completing monitoring objectives?
Resources include money, personnel, and time. There is often a balance between the number of parameters monitored and the number of events for which they can be measured that is usually driven largely by available resources. In most cases, limiting the number of parameters to those most significant to the monitoring objectives, will allow more storm events to be sampled, leading to better supported conclusions (GeoSyntec and ASCE, 2002).

Do any regulatory or legal requirements apply to the BMP or its receiving waters?
Those parameters specified in any regulatory requirements or court-ordered legal requirements must be included in the BMP monitoring program. BMP specific regulatory considerations include the use of the 80% reduction in TSS rule and the National Pollutant Discharge Elimination System (NPDES) Stormwater Phase II Regulations. The coastal zone management measures guidance require an annual average reduction in TSS loadings of 80% in construction areas (NJDA *et al.*, 2000). The U.S. EPA's Phase II requirements include provisions to track the implementation of five minimum stormwater management measures: public education and outreach, public involvement, illicit discharge detection and elimination, reduction of construction site and post-construction runoff, and pollution prevention/good housekeeping activities. The U.S. EPA included no specific requirements for chemical (or biological) monitoring information. Thus, some municipalities have developed a BMP monitoring program that compiles qualitative data only about both structural and nonstructural BMPs to satisfy their permit requirements (Hillegass, 2003). A monitoring program of this type primarily tracks contributing factors, e.g., the type of BMP used and the size of its catchment area.

Applicable surface water quality standards of the receiving water should be reviewed before the final parameter selection. For example, if the water quality criteria specify levels for total metals and the monitoring plan only calls for soluble metals or vice versa, the data may not be able to answer key questions concerning the effectiveness of the BMP.

Are existing monitoring data available?
Data from previous monitoring programs or screening studies can be useful in identifying appropriate parameters for a successful BMP monitoring program (as well as establishing estimates of the number of samples need to detect desired levels of change). Constituents prevalent in existing monitoring data, especially those above levels of concern and those targeted by the BMP should be

included, whereas those constituents rarely detected generally can be eliminated unless other circumstances exist, such as a recent change in the catchment area, or a regulatory or legal requirement (Caltrans, 2000). Selecting appropriate parameters based on existing data usually leads to a cost-effective BMP monitoring program. Screening studies to ascertain what pollutants might be of concern/interest can be valuable during the design phase of a BMP monitoring program (GeoSyntec and ASCE, 2002).

What are the prevailing land uses in the catchment area?
Different land uses have been associated with different types of pollutants and potential pollutant sources. Knowledge of the prevailing land uses in the target catchment area can help identify parameters likely to be in the wet and dry-weather runoff. For instance, if the catchment area includes a large amount of highway or industrial yard runoff, parameters such as metals, PAHs, and industry specific toxic substances are likely to be of concern (Caltrans, 2000). However, if the catchment area is heavily influenced by construction or agricultural activities, sediment pollution is typical. In catchment areas subjected to combined or sanitary sewer overflows, pathogens may be of importance.

What are the beneficial uses and impairments (if any) of the receiving water?
Beneficial uses or impairments of receiving waters are often the underlying reason behind BMP implementation. Monitoring programs can be used as verification that the BMP is fulfilling its intended purpose. It has been recognized that "in many instances the water quality problem will directly indicate what variables should be monitored" (U.S. EPA, 1996). For instance, if the BMP discharges near a public beach, pathogens or bacterial indicator monitoring will be important. Or, if the BMP discharges to a stream that supports a healthy game fish population, then in-stream biological indicators may be useful.

Are there any parameters that are particularly useful for evaluating the type of BMP being monitored?
Some parameters will be more important than others, depending on the type of BMP being monitored. For example, soil permeability may be an important parameter for infiltration BMPs, but less important for ponds, especially those that are concrete or clay lined. Alternatively, settleable solids may be an important parameter for wet ponds, but less important for infiltration BMPs. Another example is particle settling velocity distribution for BMPs that depend on removal by inertial separation of suspended solids, and particle size distribution for BMPs that depend on filtration (Field and O'Connor, 1996).

Are there any contributing factors that would be useful in interpreting data from the primary parameters selected?
Collection of supporting data can greatly enhance the usefulness of BMP monitoring data (Caltrans, 2000). The most striking example is the measurement of hardness when metals have been selected as a primary parameter, as discussed above in Section 4.2.1.1. Other examples of contributing factors are given in Section 4.2.1.5.

Are the parameters typically monitored constituents?
Whenever possible, the parameters selected should be typical (e.g., those typically found in urban stormwater runoff at levels of concern). Typically measured parameters are most likely to have straightforward and reliable methods of analysis (GeoSyntec and ASCE, 2002). Their data is easily comprehensible by those in the environmental profession, which gives them a broader usefulness. If typical parameters are measured, the data will be easily comparable to other studies with similar circumstances and will tend to be more economical overall.

4.3 Monitoring Nonstructural BMPs

According to an international survey, the use of nonstructural BMPs is on the rise and the trend is expected to continue (Taylor and Wong, 2002a). Yet, there is a paucity of high-quality studies on their effectiveness in improving the quantity and quality of stormwater. This lack of monitoring of nonstructural BMPs has been cited on more than one occasion "as a major impediment to their adoption" (Taylor and Wong, 2002a). Monitoring of nonstructural BMPs is inherently difficult for many reasons. The most significant hindrance to monitoring nonstructural BMPs is that many of them rely on behavioral change. Behavioral changes are difficult to measure for the following reasons: (1) direct observation of behavior is usually not an option for privacy reasons; (2) self-reported behaviors are often vastly different from the truth; (3) people's attitudes toward environmental issues do not often translate into actual behaviors; (4) developing suitable quality controls is difficult; and (5) there is a time lag between awareness about an issue and actual behavioral change, which could require up to ten years (Taylor and Wong, 2002a; Taylor and Wong, 2002b).

A second reason for the difficulty monitoring nonstructural BMPs is the lack of defined inflows and outflows. Therefore, the effects of nonstructural BMPs cannot be isolated for direct measurement. Taylor and Wong (2002a) conceded that monitoring the effects of some nonstructural BMPs (e.g., public education and outreach) on stormwater quality is "virtually impossible...or at best can be evaluated using [long-term] trend analysis." A third difficulty is nonstructural BMP programs are often backed by substantial resources and thus, strong pressures to report positive results may exist. This has the potential to skew the results of any monitoring that has been conducted (Taylor and Wong, 2002b). One final word of caution regarding the monitoring of nonstructural BMPs is that the effectiveness results collected may not be transferable to other communities (Taylor and Wong, 2002a). People's behaviors are shaped by social, educational, economical, and regional factors; therefore, what works in one place, may not be universally effective.

Despite its difficulties, nonstructural BMP monitoring is feasible using a number of approaches. Taylor and Wong (2002a) identified seven approaches to evaluating nonstructural BMPs. Many of the approaches identified are qualitative in nature, such as monitoring public awareness of target issues and monitoring self-reported behaviors. Obtaining quantitative monitoring data on nonstructural BMPs can be accomplished in three ways. The first is through modeling. Through the use of models, qualitative data can be manipulated into quantitative estimates. For example, the percentage of people who claim to have changed their use of lawn chemicals may be translated into an estimated reduction of herbicides in runoff. The second way is the direct measurement of loads prevented from entering the stormwater. This approach has limited applicability to BMPs such as street sweeping and catchbasin cleaning where pollutants are collected and can be weighted. The third way to obtain quantitative data for nonstructural BMPs is long-term trend monitoring of a downstream, end-of-catchment system, such as in-stream parameters of the receiving water (see watershed monitoring below), a structural BMP influent, or other end-of pipe system. Although the effects of specific BMPs cannot be isolated, data collected before and after nonstructural BMPs implementation, or using a reference watershed can produce valuable quantitative data in support of the effectiveness of nonstructural BMPs.

4.4 Watershed Monitoring

A watershed provides a contained unit of receiving waters with ecologically significant boundaries (WEF and ASCE, 1998). It will cumulate the impacts of human activities such as urbanization, which has been directly linked to receiving water degradation (U.S. EPA, 1999). The concept of best management practices and all the related techniques and technologies were developed with the ultimate goal of minimizing the impacts of urban runoff (both wet and dry) on receiving waters. Therefore, watershed monitoring may produce a more accurate measure of the true value of BMPs. For some structural BMPs with no defined inflow and outflow and for nonstructural BMPs, watershed monitoring will be the only valid monitoring option. Also, watershed monitoring has the ability to combine the effects of a wide-reaching stormwater management program. When these programs involve a number of nonstructural and structural BMPs, it can be more economical to capture the effects of all BMPs in one watershed monitoring program than to attempt to monitor each BMP individually. However, a disadvantage is that the effect of any one BMP becomes nearly impossible to ascertain. Although watershed monitoring typically refers to in-stream monitoring of receiving waters, the same techniques may be applied to pipe systems or "sewersheds" when appropriate.

Of the four BMP monitoring approaches discussed in Section 4.1, only one, the influent/effluent approach, truly isolates the BMP's effects on water quality. The remaining three effectiveness monitoring approaches, before/after, upstream/downstream, and control (or reference or paired) watershed, can be applied to the watershed as a whole. An example of the before/after monitoring approach is the Englesby Brook monitoring project being conducted by the USGS in Burlington, Vermont (Medalie, 2000). Several BMPs were scheduled to be constructed for stormwater management. An in-stream monitoring program was initiated two years before completion of the first BMP in 2002. The monitoring is planned to continue beyond completion of the final BMP for a total of seven years.

The control watershed approach is the most difficult to apply successfully because of the many variables involved (GeoSyntec and ASCE, 2002). Typically, control watersheds can differ in total area as long as the ratio of pervious to impervious areas closely approximate that of the target watershed, and other watershed characteristics also correspond relatively well. Otherwise, "as the characteristics of the two watersheds diverge...the noise in the data becomes greater than the signal" (GeoSyntec and ASCE, 2002).

Watershed monitoring also may be used for trend or compliance monitoring where sample characteristics would be compared over time or with designated-use goals or water quality standards (GeoSyntec and ASCE, 2002). A good example of this type of watershed monitoring is the U.S. EPA's Environmental Monitoring and Assessment Program (EMAP) (U.S. EPA, 2002a). EMAP uses a probability-based sampling design to obtain a statistically-valid picture of the ecological condition of statewide water bodies. Currently, planning is underway to apply the key concepts of EMAP to monitor BMPs on a watershed scale. One disadvantage to using the EMAP approach for BMP monitoring is the need for a large participating area to ensure a representative sample, which drives up the cost of implementation.

Special attention needs to be paid to measurement location in watershed monitoring, as there are no fixed sampling locations as in input/output sampling (FHWA, 2000). Sensors and flow metering devices must be placed in a location at least twenty channel widths downstream from any channel bends. The intake for automatic water sampling equipment should be located in a portion of the waterway that is relatively well-mixed in cross-section. To prevent misleading results from bedload contamination of the samples, the intake also should be fixed at a height of 100 to 200 mm (4 to 8 in.) above the streambed (FHWA, 2000). Even so, some biases can occur.

Watershed monitoring has its challenges. Because receiving waters are influenced by many other stressors (e.g., natural weather patterns), the effects of stormwater management programs may be obscured. The slow biological response to improving conditions and dilution effects that reduce sensitivity to chemical improvements may make detection of statistically significant changes impossible (GeoSyntec and ASCE, 2002). This, compounded by annual, seasonal, and even diel (Brick and Moore, 1996) variability, requires long-term continuous monitoring. A minimum of two to three years is recommended for watershed monitoring programs (FHWA, 2000). Finally, from a planning and policy point of view, watershed monitoring may not be a viable option because watersheds do not usually follow political boundaries, and involvement of all necessary municipalities may not be possible (WEF and ASCE, 1998).

4.5 Developing a BMP Effectiveness Monitoring Program

Developing a BMP monitoring program that produces useful results takes a great deal of effort before any samples are taken. The Federal Highway Administration (FHWA) (2000) produced a well presented guide to the development of a BMP monitoring program. The agency organized a BMP effectiveness monitoring program into four phases:

(1) planning phase,
(2) design phase,
(3) implementation phase, and
(4) evaluation phase.

This section follows the FHWA's organization relatively closely, but each section has been supplemented with information from additional sources. The FHWA's document is recommended for anyone involved in developing a BMP monitoring program because each phase is supported with useful examples to help conceptualize how the guidelines are put into practice.

4.5.1 The Planning Phase

The planning phase is a critical first step in developing an efficient BMP monitoring program. In the planning phase program goals are defined, background information is collected, and resources are identified (FHWA, 2000). Using this information, specific project objectives can be formulated. These objectives form the framework within which the remainder of the BMP monitoring program is designed, implemented, and evaluated. The U.S. EPA (1996) recognizes that well defined goals and objectives are "the most fundamental step in the development of a monitoring plan."

4.5.1.1 Defining Program Goals

Monitoring goals are broad statements that cover the issues of concern for a particular stormwater management monitoring program (FHWA, 2000). Examples of goals for BMP monitoring programs include, evaluation of: (1) flood control; (2) changes in runoff volume and release rates; (3) pollution reduction by a specific BMP or extended stormwater management program; (4) the possibilities of stormwater harvesting from the BMP; and/or (5) BMP longevity or maintenance requirements (Argue, 1995; FHWA, 2000; GeoSyntec and ASCE, 2002). Each goal often requires a different set of assessment methods and data needs. This section focuses primarily on concerns related to the goal of evaluating the pollution reduction of a specific structural BMP.

4.5.1.2 Collecting Background Information

Background information is essential to formulating appropriate objectives in a BMP effectiveness monitoring plan. Background information includes, site and catchment area characteristics, BMP design characteristics, receiving water reference conditions, site hydrologic assessment, and any existing data. Site and catchment characteristics and BMP design characteristics also are considered critical to evaluate efficiency differences between BMPs (Strecker *et al.*, 2000). As mentioned earlier in Section 4.2.1.5, thorough tables of these contributing factors can be found in GeoSyntec and ASCE (2002). Upstream BMPs are an important component of catchment characteristics that may affect BMP performance for certain parameters (GeoSyntec and ASCE, 2002). Receiving water reference conditions and existing data can be used as a point of comparison and for identifying areas of concern (NJDA *et al.*, 2000). A hydrologic assessment to determine a site's distribution of storm size and frequency can help to develop an appropriate temporal scale for the BMP monitoring program (FHWA, 2000). The hydrologic assessment also can determine the possibility of groundwater influences.

4.5.1.3 Identifying Project Resources

Identification of project resources is necessary to assure that the monitoring program can be attained realistically. Limitations of project resources will ultimately bound the overall scope of the program objectives. Modeling may be considered to fill in gaps when resources are short, although good field data is preferred (WEF and ASCE, 1998). Examples of areas where modeling may be used in lieu of monitoring data are the use of measured BMP effluent data to estimate receiving water concentrations instead of monitoring the receiving waters themselves and flow modeling for load determination instead of actual flow monitoring. Two additional solutions to limited resources suggested by GeoSyntec and ASCE (2002) are to break down the program goals into smaller and more manageable questions and, if possible, limit the number of parameters monitored before cutting the number of samples collected.

A more creative way to help keep costs low while gathering an extensive database on BMP effectiveness is to enlist the help of volunteers. This option has the added benefit of counting toward compliance with two of the minimum control measures of the NPDES Stormwater Phase II Regulations, public education and outreach and public participation and involvement (Baxter, 2002). Rhode Island and Connecticut currently boast successful volunteer monitoring programs. After a short training period, approximately 250 volunteers are given the supplies to perform weekly or biweekly monitoring at over 120 sites throughout the two states. In Alpharetta, GA, a similar program relies on volunteers to collect and deliver samples to the lab, but for safety reasons the volunteers do not actually handle any of the chemicals used for their water quality monitoring assessments (Baxter, 2002). The most problematic issues with employing volunteers as part of a monitoring program is quality assurance and continuity. It is imperative that a good quality assurance plan is in place and that volunteers have training on the information. However, even if these measures have been taken, adherence to the quality assurance procedures are likely to be more difficult to enforce when using volunteers. It is also more difficult to keep a monitoring program operating with regularity when relying on volunteer help.

4.5.1.4 Formulating Monitoring Objectives

Once the first three steps of the planning phase have been completed, specific monitoring objectives can be set. The FHWA (2000) suggests looking at the program goals as the question that needs to be answered and program objectives as what needs to be done to answer that question. Knowledge of background information and available resources should allow the jump to be made between program goals and program objectives. Clear and specific objectives are important, as they will drive the program design phase.

4.5.2 The Design Phase

The design phase translates the objectives into an action plan. Issues that need to be defined include monitoring approach, parameter selection, hydrologic data collection protocols, water quality data (including chemical, physical, and biological parameters) collection protocols, identification/selection of equipment and materials, and quality assurance/quality control (QA/QC) initiatives. The product of the design phase should be a quality assurance project plan (QAPP) that lays out these details, providing a pathway for meeting the monitoring program objectives. This phase is the foundation of the project and should be given considerable attention. A poorly-designed monitoring program could produce misleading data and erroneous conclusions, resulting in great deal of wasted time and money (GeoSyntec and ASCE, 2002). The following subsections are brief discussions of key issues included in the design phase.

4.5.2.1 Monitoring Approach

The four main monitoring approaches have already been discussed in Sections 4.1 and 4.4. The influent/effluent approach to BMP monitoring is used most often. It has the advantage of being a straightforward approach, isolating the effects of the BMP, usually costing substantially less than other approaches, and having a short time requirement to reach significant findings (FHWA, 2000). However, receiving water response has been indicated as "a better gauge of long-term BMP effectiveness..." (Clary *et al.*, 2001). Three so-called watershed approaches monitor the receiving water response. These approaches, upstream/downstream, before/after, and control (or reference or paired) watershed, are discussed in length in Section 4.4. Typically, the influent/effluent approach is the optimum choice, with watershed approaches reserved for BMPs that do not have a defined input and output, or for cases in which receiving water quality is part of the monitoring program objectives.

4.5.2.2 Parameter and Methods Selection

Selection of appropriate parameters is an important step. As such, parameter selection has been expanded into its own section (4.2), and therefore will not be covered here in detail. A few additional issues regarding methods selection should be addressed. First, the selected methods should be standard. Using unconventional methods makes comparisons between monitoring programs unreliable, diminishing the value of the data collected (Ruby and Kayhanian, 2003). Second, methods that require pre-acidification of sample bottles can be problematic, especially when using automatic samplers (Rushton, 2002). Experience with these methods showed that the preservatives were prone to evaporation and sample size estimation was difficult. Third, the detention limits of the methods chosen should reflect the expected ranges of the pollutants of concern. Although a data set of mostly non-detects is not completely useless, actual numbers are always more desirable.

4.5.2.3 Hydrologic and Hydraulic Data Collection

A robust BMP monitoring program begins with hydrologic monitoring. Rainfall intensity, duration, and total precipitation volume per storm can be monitored with recording precipitation gauges (Church *et al.*, 1999). Non-recording precipitation gauges are not recommended as they do not provide the intensity-duration timing necessary to obtain a clear picture of the pollutant mobilization energy of the storm. Three types of precipitation gauges are generally acceptable, weighting, float, and tipping-bucket for rainfall (Church *et al.*, 1999). Weighting gauges and heated tipping-buckets are the acceptable methods for snow, which may be an important issue for BMP monitoring studies in northern climates. Precipitation gauges are most useful when located near the water quality monitoring stations, especially when used to initiate autosamplers. Using precipitation to initiate autosamplers is somewhat problematic because of the spacial variability inherent in rainfall. Also, depending on land-use and seasonal conditions (e.g., snow event), precipitation does not always guarantee runoff. For these reasons, an increase in flow rate is usually a better trigger for autosamplers. Care should be taken so that the gauge is not under trees or other vegetation that will intercept the rainfall (GeoSyntec and ASCE, 2002). It also may be advantageous to collect a precipitation sample for water quality monitoring. This becomes especially important if nitrogen species are of concern because "much of the ammonia and nitrate in stormwater are deposited directly in rainfall" (Rushton, 1998). For BMPs with a large surface area, the input of precipitation falling directly on the surface of the BMP may be significant.

Hydraulic data is essential for accurate determination of pollutant loadings (GeoSyntec and ASCE, 2002). Concentration data alone does not convey the full impact of stormwater or BMP effluent impact on the receiving water. A complete hydraulic picture includes flow rates and volumes of BMP inflow and outflow during a storm event, as well as dry-weather flows, bypasses or overflows, and groundwater flows. Flow rates can be estimated in a number of ways, but the method recommended by GeoSyntec and ASCE (2002) for BMP monitoring programs, is through the combination of a primary control device (flume or weir) and a secondary control device (float gauge, bubbler, pressure transducer, ultrasonic level sensor, ultrasonic uplooking, radar/microwave sensor, and pressure probe). Primary control devices have channel geometries that have been calibrated against the depth of the water flowing through (flume) or over (weir) the device to develop a relationship between water depth and flow. The secondary device measures the water depth. GeoSyntec and ASCE (2002) contend that the accuracy of primary control devices makes any additional costs associated with installing the devices worthwhile because errors in flow measurements will propagate through data analysis (GeoSyntec and ASCE, 2002). Weirs are generally more accurate and easier to construct and install than flumes (GeoSyntec and ASCE, 2002). Primary devices do have some difficulties. By constricting flow, they cause water to backup, allowing for sedimentation (Church *et al.*, 1999). This will change the physical and chemical characteristics of the samples, leading to inaccurate conclusions. Flumes are less likely to cause this problem, due to their self-flushing design, and although they may be more difficult to construct and install, when done correctly, accuracies are on par with or greater than weirs (Church *et al.*, 1999). Therefore, if a primary device is to be used, Church *et al.* (1999) recommends flumes. Guidelines for selecting an appropriate secondary device can be found in GeoSyntec and ASCE (GeoSyntec and ASCE, 2002).

Regardless of how flow is measured, caution should be taken when comparing results collected using different methods. Individual flow measurement methods can have considerable variability and bias (Church *et al.*, 1999). Flow devices should be calibrated frequently to ensure measurements are as accurate as possible.

Measurement of inflow and outflow data during a storm event is typical, but dry-weather flows not associated with hydrologic monitoring can be important, as well. It had been recognized that over the long-term, dry-weather pollutant loads often dominate those resulting from wet-weather flows for many pollutant types (Pitt *et al.*, 1993). Therefore, flow records should be inspected for the frequency and volume of dry-weather flows. If dry-weather flows are a significant part of total flow through the BMP, water quality samples of dry-weather flows should be taken. Flow through bypasses or overflows must also be considered. Bypasses and overflows can significantly affect the efficiency of a BMP system as a whole (GeoSyntec and ASCE, 2002). Disregarding these flows during data analysis will result in misleading conclusions. If direct flow measurements are not possible, accurate measurements of inflow and outflow data may be used to estimate bypass or overflow volumes through mass balance calculations. Pollutant loadings, instead of concentrations, will provide more accurate efficiency evaluations in these situations.

In addition to surface hydraulics, subsurface hydraulics also may be necessary. Groundwater contributions are especially important for BMPs located at or near the groundwater table (U.S. EPA, 1999). Groundwater contributions can be directional and, thus, may require the installation of both shallow and deep piezometers to estimate horizontal and vertical flow, respectively (Reinelt and Horner, 1995). If groundwater is found to have significant inflow to the BMP, water quality samples of the groundwater should be taken. Reinelt and Honer (1995) found that 80% of the total phosphorus loading to a wetland BMP was attributable to groundwater inflow. Had groundwater sampling not been conducted, efficiency calculations would have erroneously concluded that the wetland itself was releasing phosphorus when in fact it was reducing groundwater concentrations. This example indicates the importance of comprehensive flow and pollutant budgets in BMP monitoring programs.

4.5.2.4 Water Quality Data Collection Protocols

Water quality data in this section refers to all chemical, physical, and biological parameters that are used to characterize the quality of water. This data is generally the heart of the BMP monitoring program. It has been suggested that differences in sample collection may be the largest source of variation in BMP performance and as such thoughtful design of water quality data collection protocols is imperative (Strecker, 1994). The sampling location, sampling method, sampling frequency, and sample representativeness all need to be considered in the BMP monitoring design. Each of these four elements will be discussed briefly below.

Sampling Location. Sampling equipment must be fixed in a location that is easily and safely accessible. If a primary control device is used, inflow and outflow samples should be taken a short distance upstream of the device in a well-mixed section of such flow. If automatic samplers are being used, the location of the intake tubing should not be significantly lower in elevation than the autosampler unit. It was recommended that the difference between the intake elevation and the sampler unit should be no more that 20 feet and preferably less (FHWA, 2001). Studies have found that sample volumes became unacceptably variable at high lift heights (FHWA, 2001). Also, in theory the representativeness of the collected sample with respect to larger sediments may be compromised at high lift heights. Although some research shows that this may not be as much of a problem as one may expect (FHWA, 2001). For BMP systems that are arranged in a "treatment train", sampling at intermediate locations within the system may be useful to evaluate the significance of each individual component. Some special considerations regarding selection of an in-stream sampling location are discussed in Section 4.4.

Sampling Method. Samples can be discrete, composite, or specialty. Discrete or grab samples can be taken at a single point during the storm or sequentially throughout the storm. A discrete sample taken at the onset of inflow can be used to characterize first-flush phenomenon (GeoSyntec and ASCE, 2002). Discrete samples are required for some parameters like volatile organic carbons that degrade if the sample is left exposed. Sequential discrete samples can be used to create a "pollutograph" of pollutant concentration verses flow or time (depending on how the series is weighted). A pollutograph can provide "insight into the performance of a BMP [over time or] under various hydraulic loadings" (U.S. EPA, 1999). Peak concentrations and the time during which water quality standards were exceeded (when applicable), also can be determined from discrete sequential sampling (GeoSyntec and ASCE, 2002). Synchronizing the times at which the discrete samples are pulled with flow measurements allows for accurate calculation of loading rates and flow-weighted averages (i.e., event mean concentrations).

Composite sampling can be implemented in several ways. The simplest way, constant time-constant volume composite, is not common and generally not acceptable for compliance with stormwater regulations (FHWA, 2000). Time-weighted composites are not reliable for estimation of mean storm concentrations or pollutant loads, and thus are not recommended for a BMP monitoring program (GeoSyntec and ASCE, 2002). There are three flow-weighted ways to collect composite samples that are considered suitable: (1) constant time-volume proportional to flow rate; (2) constant time-volume proportional to flow-volume increment; and (3) constant volume-time proportional to flow-volume increment (GeoSyntec and ASCE, 2002). For manual sampling, the first option is most practical. For automatic samplers the third option is preferred because it is generally more accurate than the other two flow-weighted sampling methods. (GeoSyntec and ASCE, 2002). The increased accuracy is due to the combination of continuous flow measurements (the first option is based on a single flow measurement) and the elimination of sample splitting, which could introduce contamination (the varying volumes in second option will require sample splitting).

The constant volume-time proportional sampling method adds individual aliquots to the composite at a greater frequency when flows are high. Determining the exact autosampler parameters, such as volume of aliquots and flow triggers, can be difficult. Low flows of smaller storms can lead to insufficient sample volume, whereas, larger storms may fill the composite sample bottle too quickly, truncating the later part of the storm. The resulting sample will not be representative of the entire storm, thus calculation of an event mean concentration will be invalid (Rushton, 2002). Problems may also arise when a large aliquot volume is combined with periods of very high flows. Under these circumstances the flow-weighted sampler may be triggered to collect the next sample while the previous sample is still being drawn. Thus, fixed autosampler parameters are not appropriate for all storms. As a targeted storm approaches, estimates of storm size and peak intensities should be made so that autosampler parameters may be adjusted as necessary.

Sampling Frequency. The sampling frequency is the number of dry-weather samples taken or storm events sampled. Given the great temporal and spacial variability of storm events, "a small number of samples are not likely to provide reliable indication of stormwater quality at a given site or the effect of a given BMP" (GeoSyntec and ASCE, 2002). Estimates of the sampling frequency necessary to provide a reasonable average event mean concentration for a particular location are eight to ten events (WEF and ASCE, 1998), however, statistical methods should be used to determine an appropriate number of samples. These methods require an initial estimate of sample variability, which can be obtained from existing data or from the literature, a minimum level of detectable change, and a desired confidence level and power. Typically, calculations are performed with a 95% confidence level at 80% power (GeoSyntec and ASCE, 2002). This means that there is a 5% chance of finding significant change where none exists (type I error) and a 20% chance of overlooking a significant change that does exist (type II error). The chosen minimum level of detectable change makes a large difference in the necessary sampling frequency. For example, the minimum number of samples required to detect a 5, 20, and 50% change in mean EMCs for three key parameters at two monitoring locations are shown in Table 4-1.

Table 4-1. Minimum Sampling Frequency Necessary to Detect Significant Changes in Mean Concentrations for Key Parameters at Two Sites in Portland, Oregon[1]

Site	Parameter	Minimum Number of Samples to Detect the Indicated Percent Change[2]		
		5%	20%	50%
1	TSS	202	14	4
	Phosphorus	224	16	4
	Copper	442	29	6
2	TSS	61	5	2
	Phosphorus	105	8	3
	Copper	226	15	4

[1]reproduced from (Strecker *et al.*, 2000) pg. 183
[2]80% power of detection in the mean of the EMCs

The increase in required sampling frequency between the three parameters indicates that variability in percent reductions was much greater for copper compared to TSS and phosphorus at each site. Their analysis shows that detection of a small change may take many years of sampling every storm, given that some locations in the U.S., especially in the west/southwest, experience only 10 to 20 measurable storms per year (Driscoll *et al.*, 1989). Some have estimated that at least five years are necessary to ensure important performance information is not missed (Urbonas, 2000). The time of year during which these samples are taken is also important. Due to the seasonality of many factors involved in BMP effectiveness, samples must be collected covering all seasons, yet rarely is data collected in the winter in cold regions.

Irrespective of the above discussion, for the most representative analysis samples should be taken as frequently as economically possible. Ideally, the sampling period should be continuous, covering all seasons, in dry as well as wet weather. Synchronizing the samples with flow measurements will produce accurate long-term pollutant mass reductions (Field and O'Connor, 1996).

Sample Representativeness. The NPDES permit requirements identify key characteristics that result in a representative storm. The total precipitation and duration of a representative storm should be within 50% of the average event for a given location, produce equal to or greater than 0.1 inches of precipitation, and have an antecedent dry period (i.e., less than 0.1 inches) of at least 72 hours (GeoSyntec and ASCE, 2002). The antecedent dry period will set up a worst-case scenario, allowing for build up of pollutants to be washed away with runoff. For BMP monitoring purposes outside of a NPDES permitting application, strict adherence to the representative storm is not necessary or likely not desirable. There is no truly "representative storm" and monitoring only under these guidelines may skew the results. It is preferred that monitoring be performed under a wide variety of conditions and storms (GeoSyntec and ASCE, 2002). Seasonality is another important factor of sample representativeness. Variation in human activity, temperature, and precipitation type throughout the year should be considered when determining the period over which the monitoring program is run.

During each storm event, sample representativeness can be determined using a percent capture requirement (Caltrans, 2000). Percent capture is defined as the volume during which sampling occurred divided by the total event volume. The California Department of Transportation (Caltrans) (2000) also developed a minimum acceptable number of aliquots per composite sample (when monitored on a flow-proportional basis) for an event based on the total precipitation of the event in inches. Their percent capture and minimum number aliquots requirements for a representative storm event are presented in Table 4-2. Once a sample has been collected, this table is consulted and the sample is analyzed only if it is deemed representative.

Table 4-2. Caltrans Requirements for Composite Sample Representativeness†

Total Event Precipitation (inches)	Minimum Acceptable Number of Aliquots	Percent Capture Requirement
> 0.25	6	85
0.25 - 0.50	8	80
0.50 - 1.0	10	80
< 1.0	12	75

†reproduced from (Caltrans, 2000) pg. 10-10, Table 10-1.

The minimum acceptable number of aliquots for Caltrans shown in Table 4-2 are lower than that reported elsewhere. In order to calculate an unbiased event mean concentration, WEF and ASCE (1998) estimated between 15 and 30 aliquots must be taken per storm, while GeoSyntec and ASCE (2002) estimated a lower range of between 12 and 16 aliquots per storm. No correlation with event size was made for either of their estimates.

Sampling of dry-weather flows for baseline conditions should be less variable, but seasonal and meteorological conditions should be considered. When the sampling location "is subjected to base flows, runoff related flows should be separated from low flows" by at least 10% (FHWA, 2000).

4.5.2.5 Selection of Equipment and Materials

The two major types of equipment for a monitoring program are flow equipment and automatic sampling equipment. A variety of secondary flow devices are available. GeoSyntec and ASCE (2002) describes the different devices for measuring flow depth and considerations for their use. Although not recommended, flow velocity equipment may be used with secondary devices to determine flow volume in place of primary devices. If used, electromagnetic and ultrasonic equipment for measuring flow velocity is preferred (GeoSyntec and ASCE, 2002).

Automatic samplers are usually well worth their capital expense because of the difficulty of predicting weather. Paying a manual monitoring crew to be ready to start sampling as soon as the runoff begins can result in wasted time and money (GeoSyntec and ASCE, 2002). However, automatic samplers can be a significant source of variability. Shelley and Kirkpartrick (1975) tested over 200 models of automatic samplers and found "marked differences in results obtained with different types of equipment." The type of pump used in the automatic sampler may affect the representativeness of the sample. The intake or suction velocity achieved should equal or exceed that of the flow being sampled (Shelley and Kirkpatrick, 1975). Larger sediments are especially effected as lower capacity pumps may not be able to lift the larger solids from a lower elevation intake into the sample bottles. Also, peristaltic pumps and sample transport lines with internal constrictions and sharp twists and bends may break up or prohibit passage of larger solids (Shelley and Kirkpatrick, 1975; Strecker, 1994). Because stormwater has a tendency to become stratified, especially during low flows, in low velocities, or at the effluent end of a BMP where larger and more dense particles have begun to settle out, use of a multi-leveled intake port could provide a more representative sample (Field *et al.*, 1997). Automatic sampler intake tubing should have a minimum inner diameter of 1.0 to 1.3 cm to assure adequate capture of larger particles (Shelley and Kirkpatrick, 1975). The intake tubing should be of opaque material since clear plastic tubing may be subject to algae growth from exposure to sunlight, which will contaminate the samples and may alter sample volume (Rushton, 2002). Also, steps may be necessary to ensure that the intake tubing does not freeze during the winter months in northern climates.

4.5.2.6 QA/QC Initiatives

QA/QC procedures ensure that the data collected is of reliable quality and, therefore, are an important part of every BMP monitoring program. The QA/QC procedures provide a quantitative measure of accuracy (FHWA, 2000). Field QA/QC initiatives include, field blanks, field replicates, field sample volumes (e.g., percent capture as discussed in Section 4.5.2.4), and chain of custody procedures that emphasize accurate labeling and other details such as legal issues (GeoSyntec and ASCE, 2002). Laboratory QA/QC initiatives include, method blanks, laboratory replicates, matrix spikes, and external reference standards (GeoSyntec and ASCE, 2002). Acceptable levels of variance and error should be established.

4.5.2.7 *Quality Assurance Project Plan*

The main product of the BMP monitoring program design phase is the QAPP. The QAPP lays out all the design details discussed in Sections 4.5.2.1 through 4.5.2.6. It will serve as a guide for all personnel involved. This will ensure consistency in technique between the program's team members. Thorough review, especially in terms of quality assurance, is highly recommended for a QAPP before implementation. A good source of guidance regarding the preparation of a QAPP is the U.S. EPA's *Guidance for Quality Assurance Project Plans* (U.S. EPA, 2002b).

4.5.3 The Implementation Phase

The implementation phase involves three main actions, equipment installation and testing, sample handling and processing, and preliminary review of results (FHWA, 2000). The preliminary review of results compares collected data against the QA/QC initiatives established in the QAPP and checks that the data are reasonably close to expected ranges. All results that fail the QA/QC measures should be flagged and eliminated if necessary. Operational differences between the QAPP and actual practice inevitably occur during the implementation phase. All changes to the QAPP should be documented.

4.5.4 The Evaluation Phase - Quantifying BMP Efficiency

Once a good set of quality assured data has been produced, the evaluation phase begins. The evaluation phase has one main objective, data analysis. Sample processing typically produces pollutant data in the form of concentrations. Pollutant loads can be calculated through the mathematical combination of concentration data with the associated flow data. Loads are useful information when evaluating long-term impacts (GeoSyntec and ASCE, 2002). Evaluation of BMP efficiency over time within a single storm event on the basis of concentration or loading requires the inflow and outflow data either to be matched in time or offset by the hydraulic residence time of the BMP. Neither approach is considered acceptable. Generally BMPs, even detention-type BMPs, do not act like a completely-stirred tank reactor (i.e., there is some lag between the influent and the effluent concentration) nor do they act like an ideal plug flow reactor (Law and Band, 1998). When it rains and a retention type BMP begins to produce effluent, most or all of the water leaving the BMP is the water left in the BMP from the previous storm. In fact, if a BMP was designed to hold a volume of water equal to that of the average storm, as is often the case, 60 to70% of all storms would only displace the stagnant volume without ever being a major part of the effluent (Strecker, 1994). Thus, matching influent and effluent in time is not accurate. Offsetting the inflow and outflow by the hydraulic residence time implies that the BMP acts as a perfect plug-flow reactor that is also generally not accurate (GeoSyntec and ASCE, 2002). Thus, inner storm analysis is not recommended.

Under most circumstances the most useful parameter for comparing/combining pollutant data between storms is the EMC (GeoSyntec and ASCE, 2002). Formally, the EMC is the total pollutant load of the storm event divided by the total volume of the storm event. More simply, the EMC is the concentration of a composite sample when flow-weighed sampling is performed. The usefulness of the EMC diminishes when comparing pollutant data between BMPs. In these cases, pollutant loads, not concentrations, are much more appropriate because loads are normalized by volume. Both flow-weighted composite sampling and discrete sampling with synchronized flow measurements can be used to calculate loads.

BMP pollutant removal efficiency is then calculated using any number of methods, including percent removal, summation of loads, regression of loads, reduction in mean concentration, irreducible concentration, achievable efficiency, removal relative to water quality limits, Lines of Comparative Performance©, various multi-variate and non-linear models, effluent probability method and linear regression of input versus output concentrations. The details of all but one of these methods (linear regression of input versus output concentrations, discussed in detail below) and their strengths and weaknesses are covered in GeoSyntec and ASCE (2002). The majority of the eleven methods to evaluate BMP pollutant removal efficiency listed here are rarely used or used only under specific circumstances. However, three of the methods (percent removal, effluent probability method, and linear regression of input versus output concentrations) warrant further discussion.

The most common method to evaluate the pollutant removal efficiency of a BMP is percent removal. Sometimes referred to as the efficiency ratio, percent removal is defined as the average difference between the inlet and outlet EMCs, divided by the average inlet EMC. This measure of efficiency has been faulted because: (1) BMP percent removals are typically not uniform across a wide range of influent water quality; and (2) a low influent EMC can result in a low BMP effectiveness, even if the effluent concentration is near or at the limit of detection (Clar *et al.*, 2003). Current regulations or performance standards based on percent removals are ironic in that those installations with the worst water quality are those where the percent removal standards are most often met, whereas installations in areas with good upstream source controls often produce failing percent removals despite their superior water quality (Urbonas, 2000). Therefore, summary statistics on the inlet and outlet data should always be included in reports on BMP effectiveness (Strecker *et al.*, 2000). Despite its possible pitfalls, anticipated percent removal is almost always used in selection and design of BMPs for pollution control. Therefore, percent removal is a practical method of evaluating BMP monitoring data, especially when it can be viewed along side particle settling velocity and/or size distribution for the same data set. When care is taken to perform monitoring with a mass balance approach percent removal can be a reliable measure of pollutant removal efficiency (Field and O'Connor, 1996). Long-term continuous monitoring must be performed when using such an approach to overcome the variability of influent and effluent characteristics with respect to dry- versus wet-weather flows, season, and year.

The method recommended by GeoSyntec and ASCE (2002) for evaluating BMP effectiveness is the effluent probability method. This method is statistically based. Statistical evaluation of the observed differences is important in terms of establishing the BMP had an effect given the great degree of variability inherent in stormwater monitoring (Strecker *et al.*, 2000). The average influent and effluent EMCs are examined for a statistically valid difference, then the data (usually log-transformed) are plotted on a standard parallel probability plot. Along with an indication of central tendency and variance, these plots will show useful information such as the degree of consistency of removal over the full range of observations (GeoSyntec and ASCE, 2002). When using the probability method, care should be taken in averaging the influent and effluent EMCs. The time-scale across which the data is aggregated should not be more than seasonally based, due to the seasonality of stormwater and BMP environmental characteristics (Law and Band, 1998). Such seasonal variability may mask statistically valid differences when averaged annually. A limitation of the effluent probability method is the large data requirement necessary to form a clear and useful picture of BMP performance efficiency. Presentation of standard box-and-whisker plots to summarize the log-transformed inflow and outflow data also have been recommended (Strecker *et al.*, 2000). More recently, analysis of the data in the BMP stormwater database showed the raw inflow and outflow data was much less variable that the combination of this data into percent removals (Strecker *et al.*, 2004). Average effluent water quality was suggested as the most useful descriptor of BMP efficiency.

The most recent method for evaluating BMP effectiveness is linear regression of input versus output concentrations proposed by Barrett (2004). This method pairs influent and effluent EMCs on a storm-by-storm basis. The resulting linear equation can reveal several key points about BMP performance. If the regression line is not statistically significant, then the influent water quality does not have an effect on the effluent water quality for the particular BMP and parameter in question. In this case, reporting the expected effluent water quality (i.e., average effluent EMC) will be more useful. If the regression is significant, the constant of the regression can be likened to the irreducible effluent concentration of the BMP. If the constant is near zero or negative, then 100% efficiency may be achievable for the BMP and parameter in question. The slope of the regression is an indication of percent removal. When the influent concentration is sufficiently larger than the regression constant, then the expected percent removal will be approximately one minus the slope. If the slope is greater than one, net export of the parameter can be expected. The Barrett linear regression method can also be used to assess pollutant load reductions (Barrett, 2004). Calculation of the load reductions requires water balance measurements in order to determine the fraction of influent lost to infiltration and evapotranspiration.

A possible disadvantage of the Barrett linear regression method is that variability between storms in terms of flow, intensity, antecedent periods, etc. will translate into overly variable pairings between influent and effluent concentrations. Therefore, it has been asserted that matching average influent and effluent concentrations or loadings on a storm-by-storm basis may be of limited value especially for those BMPs with permanent volumes (GeoSyntec and ASCE, 2002). However, it is undeniable that in most cases the influent concentration will have some effect on the effluent concentration. By assessing the entire data set as a whole, the Barrett linear regression method has the advantage of allowing influent/effluent pairing while accounting for its variability.

One final issue confronted during data analysis is how to handle non-detects. Non-detects are measurements that are lower than the minimum detection limit specified by a particular method. Traditionally data that fall below the limit of detection have been treated in one of four ways: (1) excluded from the analysis altogether; (2) included in the analysis as zeros; (3) included in the analysis as at the detection limit; and (4) included in the analysis as half the detection limit. Using any one of these methods will bias the results. Option (1) will overestimate the mean, while the remaining three options will underestimate the mean (Strecker, 1994). FHWA (2000) suggested using option (4) only if the percentage of non-detects is less than 15% of the data. If the percentage of non-detects is between 15 and 50%, "it is necessary to provide adjusted estimates of central tendency and dispersion that account for data below the detection limit" (FHWA, 2000). After reviewing all available techniques, Ruby and Kayhanian (2003) developed a tool that imposes a log-normal probability distribution on the detected values, then "assigns" expected values to the non-detects using regression and probability analysis. For data sets with 50 to 90% non-detects, FHWA (2000) recommends changing the statistical analysis to a percentile greater than the actual percentage of non-detects, while data sets above 90% non-detects do not provide enough useful data to perform any statistical analysis. However, even data sets with greater than 90% non-detects still provides useful information and should be reported.

4.5.5 The Presumptive Approach

A BMP monitoring program is quite complex and involves significant effort and resources, as can be seen from the above discussion. Consequently many municipalities cannot afford to implement BMP monitoring studies. Yet at the very least, compliance with such rules as the 80% TSS removal rule for BMPs must be addressed where required. The presumptive approach to monitoring uses existing monitoring data to develop an average removal percentage for similar BMP types in similar geographical areas. As long as the BMP is designed, constructed, and maintained properly, it is assumed that the average removal percentage is being equaled. NJDA *et al.* (2000) has developed such a system for BMP monitoring in the state of New Jersey. They composed a list of thirteen accepted BMPs and their expected average TSS percent removals. The 80% TSS removal rule is assumed to be met as long as a BMP or a set of BMPs has been properly installed such that the system has an expected average TSS percent removal equal to or greater than 80% according to their list. Although the presumptive approach is simple and easily applied, more good quality monitoring data is necessary before the average reported percent removals in the literature are reliable enough for use in this manner.

REFERENCES

Argue, J.R. (1995). Towards a universal stormwater management practice for arid zone residential developments. *Water Science and Technology* **31**(1): 15-24.

Barrett, M.E. (2004). Performance comparison of structural stormwater BMPs. *Water Environment Research*: awaiting publication.

Baxter, R. (2002). Grassroots stormwater programs: getting the public involved. *Stormwater* **3**(6): 74-77.

Brick, C.M. and J.N. Moore. (1996). Diel variation of trace metals in the upper Clark Fork River, Montana. *Environmental Science and Technology* **30**(6): 1953 - 1960.

California Department of Transportation (Caltrans). (2000). *Guidance manual: stormwater monitoring protocols (second edition).* CTSW-RT-00-005. Sacramento, CA. http://www.dot.ca.gov/hq/env/stormwater/special/index.htm. May 2000. Accessed on 2003.

Church, P.E., G.E. Granato, and D.W. Owens. (1999). *Basic requirements for collecting, documenting, and reporting precipitation and stormwater-flow measurements*. Report# 99-255. United States Department of the Interior, United States Geological Survey, Northborough, Massachusetts. http://www.stormwater-resources.com/library/136PLUSGSMonitoring.pdf

Clar, M., B.J. Barfield, and S. Yu. (2003). *Considerations in the design of treatment best management practices (BMPs) to improve water quality*. EPA 600/R-03/103. U.S. Environmental Protection Agency, Office of Research and Development, Cincinnati, OH. www.epa.gov/ORD/NRMRL/Pubs/600r03103/600r03103.htm October 2003.

Clary, J., J. Kelly, J. O'Brien, J. Jones, and M. Quigley. (2001). A Key Tool to help communities meet Phase II stormwater requirements. *Stormwater* **2**(2): 44-48.

Driscoll, E.D., G.E. Palhegyi, E.W. Strecker, and P.E. Shelley. (1989). *Analysis of storm event characteristics for selected rainfall gages throughout the united states - draft*. U.S. Environmental Protection Agency, Washington, DC. November 1989.

England, G. and B. Rushton. (2003). Monitoring guidelines for measuring stormwater gross pollutants, http://www.stormwater-resources.com/library.htm#BMPs, (#204PL). Stormwater News, Cocoa Beach, Florida.

Federal Highway Administration (FHWA). (2000). *Stormwater best management practices in an ultra-urban setting: selection and monitoring*. FHWA-EP-00-002. U.S. Department of Transportation, Federal Highway Administration, Washington, DC. http://www.fhwa.dot.gov/environment/ultraurb/index.htm.

Administration, F.H. (FHWA). (2001). *Guidance manual for monitoring highway runoff water quality*. FHWA-EP-01-022. U.S. Department of Transportation, Federal Highway Administration, Office of Natural Environment, Washington, DC. June 2001.

Field, R. and T.P. O'Connor. (1996). Swirl technology: enhancement of design, evaluation, and application. *Journal of Environmental Engineering* **122**(8): 741-748.

Field, R., D. Averill, T.P. O'Connor, and P. Steel. (1997). Vortex separation technology. *Water Quality Research Journal of Canada* **32**(1): 185-214.

GeoSyntec Consultants Urban Drainage and Flood Control District and Urban Water Resources Council of ASCE (GeoSyntec and ASCE). (2002). *Urban stormwater BMP performance monitoring: a guidance manual for meeting the national stormwater BMP database requirements*. EPA 821/B-02/001. U.S. Environmental Protection Agency, Office of Water, Washington, DC.

Herricks, E.E. (1996). A protocol for wet weather discharge toxicity assessment. In: *Water Environment Foundation Technical Exposition and Conference, 69th Annual*, Dallas, TX. Water Environment Foundation, Pages 13 - 24.

Hillegass, J.B. (2003). Using an indicators database to measure stromwater program effectiveness in Hampton Roads. In: *Proceedings of National Conference on Urban Stormwater: Enhancing programs at the local level*, February 17 - 20, 2003, Chicago, IL. U.S. EPA, EPA 625/C-03/003.

Kaiserli, A., D. Voutsa, and C. Samara. (2002). Phosphorus fractionation in lake sediments - Lakes Volvi and Koronia, N. Greece. *Chemosphere* **46**: 1147 - 1155.

Law, N.L. and L.E. Band. (1998). Performance of urban stormwater best management practices, http://www.unc.edu/depts/geog/them/projects/BMP.html. University of Carolina - Chapel Hill, Department of Geology, Chapel Hill, NC.

Marsalek, J. and S. Kok. (2000). Foreward - Urban stormwater management for ecosystem protection. *Water Quality Research Journal of Canada* **35**(3): 313 - 314.

Medalie, L. (2000). USGS Factsheet 114-00: Monitoring the effectiveness of urban management practices in improving water quality of Englesby Brook, Burlington, Vermont, http://water.usgs.gov/pubs/FS/FS-114-00. United States Department of the Interior, United States Geological Survey.

New Jersey Department of Agriculture, New Jersey Department of Community Affairs, New Jersey Department of Environmental Protection, and New Jersey Department of Transportation (NJDA *et al.*). (2000). *DRAFT revised manual for New Jersey: best management practices for control of nonpoint source pollution from stormwater (Fifth draft).* http://www.state.nj.us/dep/watershedmgt/bmpmanual.htm. May 3, 2000.

O'Shea, M.L. and R. Field. (1992). Detection and disinfection of pathogens in storm-generated flows. *Canadian Journal of Microbiology* **38**(4): 267 -276.

Parikh, P. (2003). A history and overview of the clean water act and its application to stormwater runoff. *National Risk Management Research Laboratory Seminar Series*, November 6, 2003, Cincinnati, OH.

Pettersson, K. (2001). Phosphorus characteristics of settling and suspended particles in Lake Erken. *The Science of the Total Enivronment* **266**: 79 - 86.

Pitt, R., M. Lalor, R. Field, D.D. Adrian, and D. Barbe. (1993). *Investigation of inappropriate pollutant entries into storm drainage systems*. EPA 600/R-92/238. U.S. Environmental Protection Agency, Office of Research and Development, Cincinnati, OH. January 1993.

Pitt, R., R. Field, M. Lalor, and M. Brown. (1995). Urban stromwater toxic pollutants: assessment, sources, and treatability. *Water Environment Research* **67**(3): 260 - 275.

Reinelt, L.E. and R.R. Horner. (1995). Pollutant removal from stromwater runoff by palustrine wetlands based on comprehensive budgets. *Ecological Engineering* **4**(2): 77-97.

Ruby, A. and M. Kayhanian. (2003). Improving stormwater monitoring. *Stormwater* **4**(6): 12 - 30.

Rushton, B. (1998). Processes that affect stormwater pollution, http://www.stormwater-resources.com/library/001PPocesses.PDF. Stormwater News, Cocoa Beach, Florida.

Rushton, B. (2002). BMP monitoring: methods and evaluations, http://www.stormwater-resources.com/library.htm#BMPs (#147PL). Stormwater News, Cocoa Beach, Florida.

Shelley, P.E. and G.A. Kirkpatrick. (1975). *An assessment of automatic sewer flow samplers*. EPA 600/2-75/065. U.S. Environmental Protection Agency, Office of Research and Development, Cinncinati, OH. December 1975.

Strecker, E. (1994). Constituents and methods for assessing BMPs. In: *Stormwater NPDES related monitoring needs*, August 7 - 12, 1994, New York, NY. The American Society of Civil Engineers, Pages 329 - 348.

Strecker, E., M.M. Quigley, and B.R. Urbonas. (2000). Determining urban stormwater BMP effectiveness. In: *National Conference on Tools for Urban Water Resources Management and Protection Proceedings*, February 7 - 10, 2000, Chicago, IL. U.S. EPA (EPA 625/R-00/001), http://www.epa.gov/ORD/WebPubs/nctuw/Strecker2.pdf.

Strecker, E.W. and B.R. Urbonas. (2001). Assessing receiving water effects of urban stormwater BMPs. In: *Linking stormwater BMP designs and performance to receving water impact mitigation*, August 19 - 24, 2001, Snowmass Village, CO. The American Society of Civil Engineers, Pages 426 - 437.

Strecker, E.W., M.M. Quigley, B. Urbonas, and J. Jones. (2004). Analyses of the expanded EPA/ASCE international BMP database and potential implications for BMP design. In: *Proceedings of the World Water and Environmental Resources Congress - Critical Transitions in Water and Environemntal Resources Management*, June 27 - July 1, 2004, Salt Lake City, UT.

Tailor, A. and T. Wong. (2002a). *Non-structural stormwater quality best management practices - a literature review of their use, value, cost, and evaluation*. EPA Victoria Technical Report 02/11. Cooperative Research Center for Catchment Hydrology, Victoria, Australia. http://www.catchment.crc.org.au/pdfs/Technical200213.pdf December 2002.

Tailor, A. and T. Wong. (2002b). *Non-structural stormwater quality best management practices - an overview of their use, value, cost, and evaluation*. EPA Victoria Technical Report 02/13. Cooperative Research Centre for Catchment Hydrology, Victoria, Australia. http://www.catchment.crc.org.au/pdfs/technical200211.pdf December 2002.

United States Environmental Protection Agency (U.S. EPA). (1996). *Monitoring guidance for determining the effectiveness of nonpoint source controls*. EPA 841/B-96/004. Office of Water, Washington, DC. September 1997.

United States Environmental Protection Agency (U.S. EPA). (1999). *Preliminary data summary of urban storm water best management practices*. EPA 821/R-99/012. Office of Water, Washington, DC. August 1999.

United States Environmental Protection Agency (U.S. EPA). (2001). *Techniques for tracking, evaluating, and reporting the implementation of nonpoint source control measures - urban*. EPA 841/B-00/007. Office of Water, Washington, DC. January 2001.

United States Environmental Protection Agency (U.S. EPA). (2002a). *Environmental monitoring and assessment program - research strategy*. EPA 620/R-02/002. Office of Research and Development, National Health and Environmental Effects Research Laboratory, Research Triangle Park, N.C. July 2002.

United States Environmental Protection Agency (U.S. EPA). (2002b). *Guidance for quality assurance project plans*. EPA/240/R-02/009. Office of Environmental Information, Washington, DC. December 2002.

Urbonas, B. (2000). Assessment of stormwater best management practice effectiveness (chapter 7). In: *Innovative urban wet-weather flow management systems*, Field, R., J.P. Heaney, and R. Pitt, (editors). Technomic Publishing Co., Inc., Lancaster, PA. Pages 255 - 300.

Water Environment Federation and American Society of Civil Engineers (WEF and ASCE). (1998). *Urban runoff quality management*. 1-57278-039-8 and 0-7844-0174-8. Water Environment Federation, Alexandria, VA.

5 Effective Use of BMPs in Stormwater Management

Bethany Madge

CONTENTS

5.0 INTRODUCTION

The effective use of BMPs in managing stormwater runoff is dependent on many variables, e.g., design, construction, selection, placement, and maintenance. Proper BMP design and construction are essential; however, BMP design issues have been discussed earlier in Chapter 3.0 and will not be repeated here. This chapter addresses proper BMP selection, placement, and maintenance, which are also essential (U.S. EPA, 1999a). Section 5.1, BMP Selection, is written from the perspective that a manager has a predetermined site on which to place or apply the most appropriate BMP. Maintenance issues are also discussed in this section. Section 5.2, BMP Placement, takes BMP selection to the next level by applying the selection process to a watershed. In this section the manager begins only with a predetermined watershed that requires protection. The BMP selection process must then be applied to all possible sites (or multiple sites simultaneously) to ascertain the best possible options to achieve watershed protection. In a more holistic management plan, the use of multiple BMPs in an integrated approach that targets multiple pollutant types and forms can improve effectiveness over individual BMPs used alone. Integrated approaches also are investigated in this chapter in Section 5.3.

The information presented is drawn from a literature review. It is intended to assist stormwater managers in making educated decisions that will best fit their situation. However, there is a great deal of variability involved in BMP applications. Even if all the guidelines and recommendations in this chapter are followed, the expected outcomes are not guaranteed. Adherence to the guidelines can only help prevent known pitfalls and identify issues that need consideration to assure the best chances for a positive result.

5.1 BMP Selection

There are a large number of BMPs from which to choose. The U.S. EPA's Menu of BMPs has identified approximately 130 individual BMPs, categorized into the six minimum requirements put forth by the National Pollutant Discharge Elimination System (NPDES) Stormwater Phase II Regulations (U.S. EPA, 2003b). New BMPs, especially in the pollution prevention and public education and participation categories, are being developed on a continual basis. In addition there are some BMPs that are a combination of individual BMPs, i.e., low-impact development techniques. Thus, selection of one or more BMPs appropriate for a particular situation may be a difficult undertaking. Given the large number of choices, elimination of inappropriate or less cost-effective BMPs through a series of sequential steps will lead to a much smaller list of the most reasonable choices from which a final decision can be made. Six BMP selection steps have been identified and are explained below in Sections 5.1.1 through 5.1.6. These steps include:

(1) regulatory considerations,
(2) site factors,
(3) stormwater quantity issues,
(4) water quality performance - pollutant removal,
(5) cost, reliability, and maintenance issues, and
(6) environmental and community acceptance factors.

Even with the steps detailed here as guidance, BMP selection is a complicated matter. Barraud *et al.* (1999) developed a computer software program, DELTANOE, to assist stormwater managers through the BMP selection process. With the input of some site-specific facts, their program can produce a list of possible solutions, or when given a favored solution, it can produce information on technical feasibility and possible scenarios of failure. As an added benefit, DELTANOE also provides guidance on design, construction, and maintenance on each proposed solution. In early 2003, the U.S. EPA also initiated a project to develop a framework for the selection of BMPs for successful stormwater management (Lai *et al.*, 2003). The integrated stormwater decision-support framework (ISMDSF) is slated to use public domain models for hydraulic, hydrologic, and water quality routing in spatial and temporal scales. The models will be combined with watershed characteristics, BMP effectiveness, and cost data to provide watershed managers with viable BMPs or combinations of BMPs to achieve the overall water quality objectives. A geographical information system (GIS) will be used as a foundation to give the ISMDSF a watershed perspective. Both structural and nonstructural BMPs are expected to be included. The estimated completion of the initial phase, development of the overall framework and the watershed component, is in early 2005 (Lai *et al.*, 2003).

5.1.1 Regulatory Considerations

In 1999, the U.S. EPA published a ruling to expand the NPDES Stormwater Phase I Regulations of 1990 (U.S. EPA, 2003c). Once only affecting a select group of industry and municipalities with populations over 100,000, the new rules (NPDES Stormwater Phase II Regulations) apply to all urbanized municipal separate storm sewer systems (MS4s), regardless of size. Coverage of industrial activities, construction sites in particular, was also expanded in the second phase. The rules require development of a stormwater management program for permitting of stormwater outfalls which are administered by the state authority through the NPDES as of December 2002 (U.S. EPA, 2000). Elements from each of six minimum control measures (see Table 5-1) must be included in the stormwater management program. Thus, BMP selection for municipalities covered by the NPDES Stormwater Phase II Regulations, must consider options in each of the six control measure areas. As mentioned above, the U.S. EPA has identified many BMPs for each category. Table 5-1 lists the six minimum control measures and possible BMP options for fulfilling each measure.

Table 5-1. BMP Options under the Six Minimum Control Measures Required by U.S. EPA's NPDES Stormwater Regulations: Phase I and II[1]

Minimum Control Measure	Best Management Practice Options
① Public Education & Outreach	▸Lawn & garden activities education ▸Water conservation practices for homeowners ▸Education on & programs for proper disposal of household hazardous wastes ▸Pet waste management education ▸Trash management education ▸Education/outreach for commercial activities ▸Outreach programs for minority & disadvantaged communities & children ▸Classroom education on stormwater ▸Stormwater educational materials ▸Low-impact development education ▸Educational displays, pamphlets, booklets, & utility stuffers ▸Using the media for community outreach on stormwater ▸Promotional giveaways to promote stormwater awareness ▸Pollution prevention education for businesses

Table 5-1. BMP Options under the Six Minimum Control Measures Required by U.S. EPA's NPDES Stormwater Regulations: Phase I and II[1]

Minimum Control Measure	Best Management Practice Options	
② Public Involvement & Participation	▸Storm drain stenciling ▸Stream cleanup & monitoring ▸Volunteer monitoring ▸Reforestation programs ▸Wetland plantings ▸Adopt-A-Stream programs ▸Rain barrel & cistern programs[2] ▸Watershed organization development ▸Stakeholder meetings ▸Attitude surveys ▸Community hotlines	
③ Illicit Discharge Detection & Elimination	▸Failing septic systems ▸Industrial/business connections ▸Recreational sewage ▸Sanitary sewer overflows ▸Identifying illicit connections ▸Wastewater connections to the storm drain system ▸Illegal dumping ▸Non-stormwater discharges	
④ Construction Site Stormwater Runoff Control	*Runoff Control* ▸Land grading ▸Permanent diversions ▸Preserving natural vegetation ▸Construction entrances ▸Check dams ▸Filter berms ▸Grass-lined channels ▸Riprap *Sediment Control* ▸Temporary diversion dikes ▸Wind fences & sand fences ▸Brush barriers ▸Silt fences ▸Sediment basins & rock dams ▸Sediment filters & sediment chambers ▸Sediment traps ▸Storm drain inlet protection *Others* ▸Turf reinforcement mats ▸Vegetative covers	*Erosion Control* ▸Chemical stabilization ▸Mulching ▸Permanent seeding ▸Sodding ▸Soil roughening ▸Geotextiles ▸Gradient terraces ▸Soil retention ▸Temporary slope drains ▸Temporary stream crossings ▸Vegetated buffers ▸Phase construction ▸Construction sequencing ▸Dust control *Good Housekeeping* ▸General construction site waste management ▸Spill prevention & control plans ▸Vehicle maintenance & washing areas ▸Contractor certification & inspector training ▸Construction reviewer ▸BMP inspections & maintenance ▸Model ordinances

Table 5-1. BMP Options under the Six Minimum Control Measures Required by U.S. EPA's NPDES Stormwater Regulations: Phase I and II[1]

Minimum Control Measure	Best Management Practice Options	
⑤ Post-Construction Stormwater Management in New Development & Redevelopment	*Structural BMPs* ▸Dry extended-detention pond ▸Wet ponds ▸Infiltration basins ▸Infiltration trenchs ▸Porous pavements ▸Bioretention ▸Sand & organic media filters ▸Stormwater wetlands ▸Grassed swales ▸Vegetated filter strips ▸Catchbasin/Catchbasin inserts ▸In-line storage ▸Manufactured products for stormwater inlet ▸Greenroofs[3] ▸Rain gardens[2] ▸Dry wells[2]	*Nonstructural BMPs* ▸Alum injection ▸Buffer zones ▸Open space design ▸Urban forestry ▸Conservation easements ▸Infrastructure planning ▸Narrower residential streets ▸Eliminating curbs & gutters ▸Green parking ▸Alternative turnarounds ▸Alternative pavers ▸BMP inspection & maintenance ▸Ordinances for post-construction runoff ▸Zoning *Others* ▸Modular treatment system ▸Dynamic vortex separators[3]
⑥ Pollution Prevention/ Good Housekeeping for Municipal Operations	*Source Controls* ▸Pet waste collection ▸Automobile maintenance ▸Vehicle washing ▸Illegal dumping control ▸Landscaping & lawn care ▸Pest control ▸Parking lot & street cleaning ▸Roadway & bridge maintenance ▸Septic system controls ▸Storm drain system cleaning ▸Alternative discharge options for chlorinated water *Materials management* ▸Alternative products ▸Hazardous materials storage ▸Road salt application & storage ▸Spill response & prevention ▸Used oil recycling ▸Materials management	*Others* ▸Airplane deicing fluid recovery system ▸Catchbasin cleaning ▸Coverings ▸Employee training ▸Flow diversion ▸Handling & disposal of residuals ▸Environmental effects from highway ice & snow removal operations ▸Internal reporting ▸Materials inventory ▸Preventative maintenance ▸Record keeping ▸Spill prevention planning ▸Stormwater contamination assessment ▸Visual inspections

[1]Taken from U.S. EPA's menu of BMPs website (U.S. EPA, 2003b)
[2]BMPs expanded out of the On-Lot Treatment listing from the U.S. EPA's menu of BMPs
[3]Additional BMPs not specifically listed in the U.S. EPA's menu of BMPs

It has been estimated that pollutant load reductions of 25 to 40% will occur when permitted NPDES stormwater management programs are fully implemented (Taylor and Wong, 2002a). However, these estimates may be overly optimistic because municipalities often lack the resources to enforce full implementation of their stormwater programs. For example, auditing of construction sites subject to stormwater controls under one municipality found an average compliance rate of 50% (Taylor and Wong, 2002a). Yet, compliance and subsequent pollutant load reductions may improve over time as public awareness increases and attitudes and behaviors change. For instance, another municipality reported a 30 to 40% compliance rate in the first few years of the stormwater management program, but 90% compliance nearly a decade later (Taylor and Wong, 2002a). Enforcement initiatives also increase compliance, but enforcement and education both require a commitment of time and money.

The total maximum daily load (TMDL) program of the Clean Water Act may also play a role in BMP selection for stormwater outfalls whose receiving waters are listed as impaired and thereby subject to a TMDL. If the stormwater outfall is already regulated under a NPDES permit and a TMDL is required for its receiving water, then a wasteload allocation must be assigned to the outfall. The BMP or set of BMPs selected must then be able to achieve effluent loadings that meet the wasteload allocated. Even if a stormwater outfall is not subject to a NPDES permit, it may still be addressed in a TMDL under the load allocation component if it's receiving water requires a TMDL. See Wayland and Hanlon (2002) for more information on BMPs and how they fit into the TMDL program.

The Coastal Zone Act Reauthorization Amendments (CZARA), Section 6217, is another regulatory consideration. This section of CZARA outlines additional stormwater requirements for coastal states. The part of CZARA that is the most influential is the new development measure. This measure requires either post-construction total suspended solids (TSS) loadings remain at pre-construction levels or at least an 80% reduction in annual TSS loading of storm runoff to the receiving water must be achieved after the construction is over and the site has been allowed time to stabilize (U.S. EPA, 2002a). For coastal areas this one requirement may be the primary driver for BMP selection. Only a few structural BMPs listed under the 5th minimum control measure in Table 5-1 are capable of consistently achieving 80% TSS reduction individually (NJDA *et al.*, 2000). However, the 80% removal may be reached by using more than one BMP per site. More information about TSS removal efficiency is presented in Section 5.1.4.

The Endangered Species Act is a fourth federal regulation that may affect BMP selection. Waterways designated as critical habitat areas of endangered animals may require extra levels of protection by BMPs (Clar *et al.*, 2003). For example, cold water species protection may rule out BMPs with unvegetated permanent pools that are likely to contribute to temperature pollution, unless the BMP can be designed offline with significant shading over open pool areas (U.S. EPA, 2002a). Other examples include protecting existing hydrology for both fish and amphibians.

State and local municipal regulations will also need to be reviewed before final BMP selection.

5.1.2 Site Factors

Site suitability is one of the key factors to successful BMP performance, especially for structural BMPs (U.S. EPA, 1999a). Thus, particular attention must be paid to site suitability in BMP selection. Site factors such as drainage area characteristics, climate and meteorological characteristics, and physical factors at the location of BMP installation need to be considered during BMP selection. The information presented in this section as well as the remaining 5.1 subsections relates primarily to the more common structural BMPs listed in Table 5-1 (namely those BMPs covered in Chapter 3).

Drainage area characteristics include watershed size, land use (existing, at full build-out, and other planned intermediate land use changes), and proximity to sensitive receiving waters. Land use is often used as a selection factor under the assumption that a successful BMP in one drainage area will be successful in a second area with similar land uses. Studies have shown a correlation between land use and stormwater quality (GeoSyntec and ASCE, 2002; Selvakumar and Borst, 2003). However, GeoSyntec and ASCE (2002) cautions that over reliance on this correlation may not yield the desired result due to considerable variation in stormwater quality and BMP performance even within the same land use category. Whether the BMP will serve new development or be installed as a retrofit is important. For example, certain BMPs such as swales typically are more practical for a new site that does not already have a curb and gutter system. The overall directly-connected percent imperviousness of the drainage area can make a large difference in BMP selection. Ultra-urban sites with little or no pervious areas are particularly difficult to retrofit. BMPs such as disconnecting roof leaders may only lead to flooding of streets and basements if not planned carefully. Limited available land in ultra-urban sites generally eliminates most structural BMPs with large land requirements, (e.g., retention and detention ponds). Pocket retention or detention ponds, or pocket wetlands are design variations of their respective larger BMPs that may be used when land is limited. Pocket variety BMPs require less of a footprint, but serve a much smaller drainage area. Sand filtration systems due to their compact size and structured nature have been increasingly used in ultra-urban areas (U.S. EPA, 1999a). In fact, filters can be installed as underground devices which keep their footprints to a minimum. A summary of drainage area characteristics as they apply to the most commonly used BMPs is presented in Table 5-2. This table represents generalized information that may not be applicable in all situations.

Table 5-2. Generalized BMP Suitability for Relevant Drainage Area Characteristics[1]

BMP Type	Drainage Area Size[2]	Land Use	New Development or Retrofit	High Percent Imperviousness Areas	Land Area Requirements
Dry extended-detention pond	10-acre minimum	widely applicable	both, depends on land availability	yes, depends on land availability	substantial, moderate for pocket dry ponds
Wet retention pond	20 - 25-acre minimum, 5-acre maximum for pocket wet ponds	widely applicable, ideal for parking lots	both, depends on land availability, pocket wet ponds may be more suitable for retrofit situations	yes, depends on land availability	substantial, moderate for pocket wet ponds
Infiltration trench	5-acre maximum	not appropriate for construction areas or other areas with a high solids source which may cause premature clogging	new development only, soil compaction probable and cost prohibitive in existing developments	sometimes, soil compaction under trench possible and infiltrated water may interfere with existing infrastructure	minimal
Porous pavement	maximum ratio of contributing drainage area to pavement area is 3:1	parking lots, driveways, residential roads with no heavy vehicles and low traffic volume	both, retrofit of large areas may become cost prohibitive	yes, distributive nature of porous pavement may cope with low infiltration rates, and subsurface drainage systems may be installed when infiltration isn't sufficient without incresing land requirements	not applicable, product substitution with no additional space requirements

Table 5-2. Generalized BMP Suitability for Relevant Drainage Area Characteristics[1]

BMP Type	Drainage Area Size[2]	Land Use	New Development or Retrofit	High Percent Imperviousness Areas	Land Area Requirements
Bioretention	2 - 5-acre maximum	ideal for parking lot islands, and between roads and sidewalks, also good for rooftop runoff	both, retrofit of large areas can be cost prohibitive	yes, ideally suited	minimal
Sand or organic media filter	10-acre maximum for surface filters, 2-acre maximum for perimeter or underground filters	widely applicable, ideal for parking lots	both	yes, ideally suited	minimal to moderate for surface filters, minimal for perimeter or underground filters
Stormwater wetland	20 - 25-acre minimum, 5-acre maximum for pocket wetlands	widely applicable, ideal for parking lots	both, depends on land availability, pocket wetlands may be more suitable for retrofit situations	yes, depends on land availability	substantial, moderate for pocket wetlands
Grassed swale	5-acre maximum	ideal for roadsides (residential and highways) and other areas with sheet flow runoff	both, although not usually practical for retrofit in existing curb and gutter areas	no, moderately sized and decentralized land requirements are usually not available	moderate, but variable

Table 5-2. Generalized BMP Suitability for Relevant Drainage Area Characteristics[1]

BMP Type	Drainage Area Size[2]	Land Use	New Development or Retrofit	High Percent Imperviousness Areas	Land Area Requirements
Vegetated filter strip	depends on design, approx. 1 acre per 580-ft strip length	ideal for roadsides (residential and highways) and other areas with sheet flow runoff, practical for very small parking lots only	both, depends on land availability	no, impractical due to large land requirements relative to the treated drainage area	moderate

[1]Source of data: (NJDA *et al.*, 2000; NYSDEC, 2001; U.S. EPA, 2002a; Clar *et al.*, 2003; U.S. EPA, 2003b; ARC, 2001)
[2]The numbers given are approximate. Local situations may require larger or allow smaller areas.

Climate and meteorological characteristics that should be considered in BMP selection include rainfall frequency, duration, intensity, climate (e.g., arid or cold), and evapotranspiration potential. Infiltration BMPs such as infiltration trenches and filter strips may not have the capacity to handle frequent (e.g., back-to-back) and/or intense storms. Cold climates with sub-freezing temperatures at least part of the year, may present difficulty, although design modifications can usually be applied (GeoSyntec and ASCE, 2002). Arid climates with high average temperatures also may be problematic for certain BMPs. The performance of BMPs which rely on availability of water, such as vegetative systems (e.g., filter strips and swales) and those BMPs with open water (e.g., wetlands and retention ponds) can decrease without an adequate water supply. Under such conditions, they may even become a public nuisance (GeoSyntec and ASCE, 2002). Climatic considerations for most common structural BMPs are presented in Table 5-3. Evapotranspiration potential is a related issue. Frost (2003) suggests plotting rainfall versus evapotranspiration potential on a monthly basis. Areas in which evapotranspiration exceeds rainfall for a predominant amount of the year will be good sites for infiltration BMPs and nonstructural BMPs such as disconnecting impervious areas. The soils in these areas will dry out quickly and thus will be able to absorb more rainfall.

Finally, physical characteristics of the BMP site location are important factors in BMP selection. Soil characteristics such as permeability and erosiveness are some obvious examples. Infiltration systems necessitate infiltration rates of 0.5 in./h or above (GeoSyntec and ASCE, 2002). Infiltration systems also typically require at least 4 feet of soil, thus depth to bedrock or groundwater is important. If groundwater is near the surface, the possibility for groundwater contamination must be considered before selection of an infiltrating BMP. Site slope is another factor. For example, gradients for swales "should be as close to zero as possible and should not exceed 5%" (Frost, 2003). The specific geological feature, karst geology, is not appropriate for many BMPs as the formation of sink holes are likely (NJDA *et al.*, 2000). Finally, the hydraulic head of a particular site, or the difference in elevation between the drainage area and the BMP site itself is important as most BMPs operate by gravity flow through the system (NJDA *et al.*, 2000). Some structural BMPs are not affected by physical location. These BMPs are entirely manmade and self contained systems (e.g., in-line storage, catchbasins and catchbasin inserts, manufactured products for stormwater inlets, and rain barrels and cisterns). Along with the climatic considerations, Table 5-3 also summarizes physical site considerations for most common structural BMPs. As before, the recommendations/limitations listed in Table 5-3 are generalized. Local conditions and engineering requirements may call for more or less restrictions.

Table 5-3. Generalized BMP Suitability for Relevant Climate and Physical Site Characteristics[1]

BMP Type	Climate	Soils	Depth to Bedrock/ Water Table	Slope	Karst Geology	Hydraulic Head
Dry extended-detention pond	modifications necessary for both cold and arid climates	suitable for all soils, highly erodible soils may require lined low flow channel	high levels of contamination require at least 2 - 4 ft to water table, otherwise can be less	maximum of 15% upstream, locally relatively flat	impermeable liner required	6 - 8 ft
Wet retention pond	may not be suitable for arid climates where permanent pool may be difficult to maintain, design variation needed in cold climates	suitable for all soils, highly permeable soils may require an impermeable liner to maintain permanent pool	pond and ground water can intersect unless high levels of contamination expected, then at least 2 - 4 ft to water table	maximum of 15% upstream, locally relatively flat	impermeable liner required	6 - 8 ft, 4 ft for pocket wet ponds
Infiltration trench	modifications necessary for both cold and arid climates	maximum of 20% clay and 40% silt/clay content, infiltration rates between 0.5 and 3 in./h	2 - 5 ft minimum from the bottom of the trench to bedrock or the seasonally high water table, at least 100 ft from down-gradient drinking wells	maximum of 15% upstream, locally relatively flat	not practical due to sink hole formation	1 - 4 ft

Table 5-3. Generalized BMP Suitability for Relevant Climate and Physical Site Characteristics[1]

BMP Type	Climate	Soils	Depth to Bedrock/ Water Table	Slope	Karst Geology	Hydraulic Head
Porous pavement	not suitable for cold climates where sand or salt deicing is used or where pavement is subject to compaction and wear and tear from heavy snow plows, freezing of infiltrating runoff may cause frost heave if base of stone reservoir is not below frost line	maximum of 20% clay and 40% silt/clay content, infiltration rates between 0.5 and 3 in./h, otherwise use with under pavement storage	2 - 5 ft minimum from the bottom of the basin to bedrock or the seasonally high water table, at least 100 ft from drinking wells	completely flat	not practical due to sink hole formation	not applicable
Bioretention	minor modifications may be necessary for both cold and arid climates	suitable for all soils, water percolates through manmade bed	2 ft minimum to water table	maximum of 5% for both upstream and local	impermeable liner required	2 - 5 ft
Sand or organic media filter	minor modifications may be necessary for cold climates, limited application in arid climates where need to irrigate may outweigh benefit	suitable for all soils, water percolates through manmade bed	2 ft minimum to water table	maximum of 6 - 10% upstream, locally completely flat	not an issue, filters are generally concrete lined	5 - 8 ft for surface filters, 2 ft minimum for perimeter filters

Table 5-3. Generalized BMP Suitability for Relevant Climate and Physical Site Characteristics[1]

BMP Type	Climate	Soils	Depth to Bedrock/ Water Table	Slope	Karst Geology	Hydraulic Head
Stormwater wetland	not suitable for arid climates where permanent pool and vegetation may be difficult to maintain, design variation needed in cold climates	suitable for all soils, highly permeable soils may require an impermeable liner to maintain permanent pool and/or wetland soil conditions	pond and ground water can intersect unless high levels of contamination expected, then at least 2 - 4 ft to water table	maximum of 8 - 15% upstream, locally relatively flat	impermeable liner required	3 - 5 ft, 2 - 3 ft for pocket wetlands
Grassed swale	limited application in arid climates where need to irrigate may outweigh benefit	suitable for most soils with some restrictions on the most impermeable soils, or poor soils which cannot sustain healthy vegetative cover, highly erodible soils will require a flatter slope and a highly stabilizing grass variety to reduce flow velocity	usually 2 ft minimum to water table, except in wet swale design variation that intersects with groundwater	1 - 2% (4 - 6% maximum) along the length of the swale, can use check dams to adjust slopes	impermeable liner required	1 ft

Table 5-3. Generalized BMP Suitability for Relevant Climate and Physical Site Characteristics[1]

BMP Type	Climate	Soils	Depth to Bedrock/ Water Table	Slope	Karst Geology	Hydraulic Head
Vegetated filter strip	limited application in arid climates where need to irrigate may outweigh benefit	not suitable for soils with high clay content or otherwise low permeable soils, or poor soils which cannot sustain healthy vegetative cover	2 - 4 ft minimum to water table to ensure filter strip does not remain wet	2 - 10% locally, < 2% will encourage ponding, > 6% encourages concentrated flow	not practical due to sink hole formation	NA

[1]Source of data: (NJDA *et al.*, 2000; NYSDEC, 2001; U.S. EPA, 2002a; Clar *et al.*, 2003; U.S. EPA, 2003b)

5.1.3 Stormwater Quantity Issues

Stormwater quantity issues include quantity (flood and drainage) control, peak flow reduction, groundwater recharge, and water reuse. Stormwater quantity control is the historic purpose of BMPs, yet it is successfully achieved by very few. Therefore, if stormwater quantity control is an objective, BMP selection will be more limited in options. Flood reduction has been the primary purpose of extended-detention ponds, or dry ponds (U.S. EPA, 2003a). Retention ponds and treatment wetlands also provide stormwater quantity control (NJDA *et al.*, 2000). However, since these BMPs are usually large regional facilities with large catchment areas they should be designed for full build-out conditions. Otherwise, the storage area available for larger storm flows may become too small which will eventually lead to increased downstream flooding. Other BMPs that provide peak flow reduction, such as in-line storage, bioretention, infiltration systems, swales, and greenroofs may help reduce flooding, but cannot be relied on as the sole mechanism for stormwater quantity control in areas where flooding is problematic. However, peak flow reduction itself is an important factor. The excess energy contained in peak runoff flows accelerates stream bed erosion and degradation of habitat in the receiving water (U.S. EPA, 1999a). This excess energy can also mobilize pollutants leading to decreased water quality. Both the maximum flows reached and the time that flows remain elevated above normal levels are important in peak flow reduction.

In some areas groundwater recharge is an objective of BMPs. Recharge has a dual benefit. In addition to recharging essential underground aquifers, recharge also reduces stormwater runoff volume, reducing runoff pollutant loads, and thereby improving BMP effectiveness in preserving surface water quality. As with quantity control, an objective of groundwater recharge has a limited number of possible BMPs that will be effective at achieving this objective. Infiltration systems such as infiltration basins, infiltration trenches, and porous pavement are obvious choices. Less obvious choices for groundwater recharge include vegetated filter strips, swales, dry wells, and rain gardens (NJDA *et al.*, 2000). A review of the U.S. EPA/ASCE's BMP database found a reduction of stormwater volume of almost 30 percent for vegetative filter strips and 30 percent for dry detention ponds (Strecker *et al.*, 2004). These reductions are likely to directly contribute to groundwater recharge. Other BMPs analyzed (wet retention ponds, hydrodynamic devices, and wetlands) were not found to contribute significantly to groundwater recharge (i.e., runoff volumes were not reduced by these BMPs) (Strecker *et al.*, 2004).

In areas where freshwater sources are scarce, using stormwater as a water source for direct use can be a great benefit. The state of Florida, for example, is actively encouraging stormwater reuse (FDEP, 2003). Stormwater runoff can be collected and reused to meet irrigation needs for land uses, such as commercial areas and golf courses, or used for toilet flushing or industrial process waters. The water reuse option is generally reserved for retention pond type BMPs for large scale operations (U.S. EPA, 2003a). Cisterns act in a similar fashion, as do rain barrels albeit on a much smaller scale. In addition to the savings associated with using stormwater as a resource, "stormwater reuse can help to maintain a more natural, pre-development hydrologic balance in the watershed" (U.S. EPA, 1999a). Hydrologic balance is achieved through irrigation by releasing the stormwater slowly and allowing evapotranspiration, infiltration, and groundwater recharge, thereby restoring area aquifers to their natural levels.

Table 5-4 contains a summary of the ability of common structural BMPs to address the four issues of stormwater quantity. Two structural BMPs (catchbasins and catchbasin inserts, and manufactured products for stormwater inlets) have no influence on quantity issues.

Table 5-4. Generalized BMP Suitability for Water Quantity Control Issues[1]

BMP Type	Flood Control	Peak Flow Reduction	Groundwater Recharge	Water Reuse Options
Dry extended-detention pond	high, ideally suited, ~30% reduction in total runoff volume but may increase downstream flooding in some cases	high, with minor design modifications, low for common design practice of 2-yr storm control	low to moderate, depends on local soil permeability and possible basin liner or low flow channel, highest when pond is completely inundated	no
Wet retention pond	high, ideally suited, but may increase downstream flooding in some cases	high, with minor design modifications, lower for common design practice of 2-yr storm control	none to low, if local soil permeability is high enough for significant recharge then the basin must be lined inhibiting recharge	yes, ideally suited
Infiltration trench	moderate	moderate, when successfully designed and maintained all runoff flow will be diverted to groundwater	high, ideally suited	no
Porous pavement	variable, low to moderate for monolithic, moderate to high for modular	high, when successfully designed and maintained all runoff flow will be diverted to groundwater	high, ideally suited when sited appropriately, approximately 70 - 80% of annual rain falling directly onto porous pavement surface area may exfiltrate to groundwater	no
Bioretention	low to moderate, ~40% volume reduction	low on an individual basis, scale generally too small to have a significant impact	none, runoff collected in an underdrain system	possible, filtered water collected in underdrain system could be diverted to a cistern or rain barrel

Table 5-4. Generalized BMP Suitability for Water Quantity Control Issues[1]

BMP Type	Flood Control	Peak Flow Reduction	Groundwater Recharge	Water Reuse Options
Sand or organic media filter	none to low	none to low	none, runoff collected in an underdrain system, exfiltration design modifications can be incorporated to provide some recharge	possible, filtered water collected in underdrain system could be diverted to a cistern or rain barrel
Stormwater wetland	moderate to high, may increase downstream flooding in some cases; large fluctuations in water level normally associated with flood control may permanently damage wetland function	high, with minor design modifications, low for common design practice of 2-yr storm control	none, may be some immediately following installation, but build-up of sediment at the bottom will eventually stop infiltration	possible, but support of wetlands vegetation necessitates a minimum acceptable permanent volume
Grassed swale	low to moderate, reported reduction in total runoff volumes = 6 - 30% over curb and gutter system, and ~47% influent end versus effluent end	low, 2 - 6% reduction over conventional curb and gutter system	low to moderate when properly constructed and maintained; none for wet swales	no
Vegetated filter strip	low to moderate, ~30 - 40% reduction in total runoff volume (influent end versus effluent end)	low	moderate when properly maintained	no

[1]Source of data: (Schueler, 1987; WEF and ASCE, 1998; COV, 1999; NJDA *et al.*, 2000; U.S. EPA, 2002a; Clar *et al.*, 2003; U.S. EPA, 2003b; Barrett, 2004a; Strecker *et al.*, 2004)

5.1.4 Water Quality Performance - Pollutant Removal

Pollutant removal has become one of the main objectives for using BMPs. Since NPDES began permitting point sources of pollution under the Clean Water Act, the nation's waters have improved significantly (U.S. EPA, 2000). However, as point source controls improved, diffuse sources (especially, stormwater runoff), have increased in relative significance and are now a leading contributor to receiving water impairment. In particular, urbanized areas "export large quantities of pollutants during storm events" (U.S. EPA, 1999a). The NPDES Stormwater Regulations (Phase I and II) and CZARA were passed to address this problem. The NPDES Stormwater Regulations require pollutant reduction to the maximum extent practicable, while CZARA makes use of TSS as a pollutant indicator requiring 80% reduction in TSS whenever annual TSS loadings cannot be maintained at or below pre-development levels (U.S. EPA, 1999a; U.S. EPA, 2000). These regulatory initiatives and the use of BMPs in TMDL requirements have been the main drivers behind the focus on BMPs as pollutant removers.

Many studies of the pollutant removal efficiencies of BMPs can be found in the literature. The majority of these data are presented in percent removals, but observed effluent quality ranges are also sometimes reported. It must be stressed that due to the variable nature of BMPs, numerically reported efficiencies are not easily transferable between sites, years, or even storms at the same BMP. Therefore, selecting a BMP based on reported removal numbers may not produce the desired result. Literature values for BMP pollutant removal efficiency should be considered general estimates and too much emphasis on these numbers could be misleading. Pollutant removal efficiency is site specific and highly variable between storm events even within the same site. To achieve the desired water quality performance, it will be more advantageous to develop an understanding of factors that are known to contribute to BMP variability instead of focusing on specific numbers. Thus, the first portion of this section will explore the factors that affect variability in BMP water quality performance. Only after these factors have been discussed will approximate numerical efficiencies be presented. These first two subsections will focus primarily on structural BMPs. The water quality of nonstructural BMPs will be covered in the final subsection.

5.1.4.1 Factors Affecting Variability in Pollutant Removal by Structural BMPs

The factors that affect variability in structural BMP water quality performance can be divided into two main categories: (1) factors that affect perceived or measured performance; and (2) factors that affect true or actual performance. Factors affecting perceived performance are governed by how the samples were taken, measured, and analyzed. These are the elements that must be considered when developing or reviewing a BMP monitoring plan. Some major points are reviewed here briefly. Chapter 4, BMP Monitoring should be consulted for more detail. Factors affecting true or actual performance will be discussed here in more depth.

(1) Factors That Affect Perceived or Measured Performance

Many parameters must be satisfied to collect a true representative sample. A few examples of these parameters are accurate flow measurement (and synchronization with sampling time for discretes) to ensure accurate flow-weighted sampling, a large enough sampling period to capture the entire storm/runoff event and identify the presence of dry weather flows, and proper selection and placement of sampling equipment and intake lines. Once a sample is taken, analytical and human error during pollutant measurement may introduce additional variability. Finally, once the data has been generated, choices made during data analysis can have a significant impact on the final reported pollutant removal efficiencies. There are many different methods available to evaluate BMP pollutant removal efficiency. In one case study, four different methods of evaluating pollutant removal efficiency were compared using the same set of data. The four methods were percent removal, summation of loads, regression of loads, and average percent removal of individual storms (GeoSyntec and ASCE, 2002). The calculated efficiency varied greatly from -2 to 82%. When comparing across BMP types, the method chosen to analyze the data can skew the overall results. For instance, methods used to evaluate BMPs in which infiltration is the dominant pollutant removal mechanism should be based on loads, not concentrations. Comparing the pollutant removal efficiency of an infiltration basin to a wet pond on the basis of concentrations will likely favor the wet pond, even if the infiltration basin is resulting in less pollutant loading to the receiving water.

Another factor that will alter the perceived performance is whether bypass volumes are included in efficiency calculations. If a BMP is designed with a bypass or to treat only a portion of the total runoff volume, whether or not the efficiency is calculated based on the whole system (i.e., entire runoff volume) or only on the portion of runoff actually captured and treated by the BMP, will make a large difference in the reported numbers. In the past it has not always been clear in these situations which method of data analysis was used.

(2) Factors Affecting True or Actual Performance

In addition to variability in measured performance, the true pollutant removal efficiency is highly variable among sites utilizing the same BMP type, and even among storms within one BMP. Four governing factors have been identified as the key elements in pollutant removal which produce this variability: (1) active pollutant removal mechanisms; (2) BMP design characteristics; (3) influent pollutant properties and concentrations; and (4) conditions within the BMP. The components within each factor sometimes overlap or interact with each other resulting in a complexity that is extremely difficult to model with simplistic tools like percent removals. Thus, pooling of water quality performance data often results in meaningless numbers.

Active Pollutant Removal Mechanisms

As discussed in Chapter 2, mechanisms for pollutant removal in BMPs include sedimentation, filtration, infiltration, sorption, phytoremediation, biological uptake, biological conversion, floatation, and natural degradation (e.g., photolysis, hydrolysis, or volatilization) (U.S. EPA, 1999a). Each mechanism provides a different efficiency of removal for different types of pollutants. For example, sedimentation typically will provide good removal for TSS, but will not remove dissolved pollutants. The actual removal efficiency of an active pollutant removal mechanism is dependent on all three of the remaining governing factors. As shown in Chapter 2, Table 2-4, most structural BMPs incorporate more than one of these mechanisms. Thus, a particular type of BMP will be appropriate for certain pollutants depending upon the dominant removal mechanisms.

BMP Design Characteristics

The basic design of each type of structural BMP has many variations. Many of these variations can affect the pollutant removal capacity of the BMP. The following list provides some examples of prominent BMP design features that affect pollutant removal. This list is not meant to be an exhaustive coverage of all necessary elements in BMP design. BMP design is discussed in detail in Chapter 3.

- In pond (both detention and retention) and wetland type BMPs residence time is a key. It will dictate how much sediment will settle out of the water column before the effluent is discharged.
- Length to width ratio, depth, and total storage volume will also be important for sedimentation efficiency in pond and wetland systems.
- For BMPs with permanent pools, the volume of the permanent pool in relation to the area's average storm rainfall volume will determine the amount of treatment afforded. Larger pool volumes will produce better and less variable effluent quality (Barrett, 2004b; Strecker *et al.*, 2004).
- Total length and slopes are keys for a swale or vegetated filter strip. A longer length or lower slope will promote a greater level of soil infiltration and filtration by the vegetation within the BMP.
- The presence, condition, and type of vegetation will affect the efficiency of phytoremediation.
- In a sand filter, the grain size of sand will determine the filter pore size, which in-turn will affect the efficiency of filtration.
- The type of media used in a media filter will affect the efficiency of sorption and filtration.

Influent Pollutant Properties and Concentrations

The properties and concentrations of the influent pollutants to a stormwater BMP will affect the efficiency of their removal. For example, if the size distribution of solids in the stormwater influent is predominantly on the small end, then sedimentation and size exclusion filtration (as found in sand filters) will be less efficient than if the size distribution is predominantly on the larger end. Another example is the percentage of particle association of pollutants such as metals and nutrients. For influents with a high percentage of particle association, more pollutant removal mechanisms will be involved resulting in a higher overall removal efficiency. However, for influents with little particle association, the efficiency of phytoremediation, sorption, and other pollutant removal mechanisms that are more effective for dissolved constituents will govern the overall pollutant removal.

Influent pollutant properties and concentrations are determined by three very important, but secondary factors: (1) drainage area characteristics; (2) storm characteristics; and (3) climatological factors. Drainage area characteristics include land use, soil characteristics, site geology, and site topography. A different suite of pollutants will be expected depending upon the contributing land uses present. For example, runoff from high density highways will likely contain relatively high levels of metals (e.g., lead (Pb), zinc (Zn), iron, copper (Cu), cadmium, chromium, nickel (Ni), and manganese) due to normal operation and frictional wear of vehicles (FHWA, 2000). The characteristics of the soil, such as infiltration rates and sorptive capacity can influence the concentration (by reducing the total runoff volume) and the particle association of pollutants, respectively. The underlying geology of the area will dictate the characteristics of the soil. Finally, site topography can enhance the erosional energy associated with stormwater runoff. In areas with steeply sloping terrain, runoff will reach greater velocities and therefore be more likely to erode and carry larger soil particles.

Storm characteristics include intensity and duration of the rainfall, the length of antecedent dry period, and the characteristics of the rainfall itself. Storm intensity and duration will impact the energy of the runoff, thereby effecting its pollutant carrying capacity. Storm intensity times duration, i.e. total storm volume, will affect pollutant concentrations through dilution. Rainfall duration (along with the antecedent dry period which will dictate how close the soil moisture content is to it's saturation point) will also affect the amount of infiltration attainable. This is especially important in infiltrating BMPs such as infiltration trenches. The length of the antecedent dry period, up to a point, can affect the influent pollutant concentrations due to a phenomenon called build-up/wash-off (WEF and ASCE, 1998). In between storms, processes such as atmospheric deposition, wind erosion, and vehicle leaking or wear of mechanical parts will accumulate pollutants on land and street surfaces. Rainfall runoff will wash some portion of the accumulated pollutants away. The longer the antecedent dry period, the more time pollutants have to accumulate, and thus the higher pollutant concentrations are likely to be in the runoff. However, wind erosion may also remove pollutants creating a maximum build-up level. Sometimes, the rainfall itself can contribute to the pollutant loading. This is especially true if nitrogen is a target pollutant as rainfall is known to contain high ammonia and nitrate levels (Rushton, 1998).

Climatological factors include temperature, solar radiation, season, and wind. Temperature will effect the influent pollutant properties by altering all kinetic reaction rates involved. Atmospheric deposition and adsoption/desorption are examples of kinetic processes that may be affected by changes in temperature. Solar radiation, especially on unshaded, dark roofing material and blacktop paved surfaces, can increase the temperature of the runoff. Temperature itself is considered a pollutant in the Clean Water Act (Parikh, 2003). Seasonal effects on influent pollutant properties include the additional loading of de-icing chemicals, salts, and sand used during the winter. These constituents will be present in abundance during snow melt events. Influent total phosphorus concentration have been reported to be lower during the winter months (Barrett, 2004b). Large loadings of organic matter, primarily decomposing leaves will contribute high biological oxygen demand in the fall. Nutrient loadings may be seasonally high during the spring and summer months, particularly if the catchment area is fertilized. Finally, wind, as mentioned earlier, can contribute to soil erosion during dry times and also affects the patterns of atmospheric deposition.

Conditions within the BMP

Conditions within the BMP that affect pollutant removal efficiency include dissolved oxygen levels, sediment cation exchange capacity, thermal stratification, BMP vegetation characteristics, climatological factors, possible groundwater intrusion, and BMP age and regularity of maintenance. Dissolved oxygen levels, in addition to being significant for oxygen-dependent aquatic organisms, also may control the concentrations of certain pollutants within a retention (wet) type BMP. Low dissolved oxygen concentrations may cause phosphorus and the metals cadmium, copper, iron, and zinc, to leach out of bottom sediments (Rushton, 1998). The cation exchange capacity of the sediments of wetland and wet pond BMPs will affect the efficiency of sorption as a pollution removal mechanism. Thermal stratification within a BMP with a permanent pool will affect settleability of solids due to the establishment of density gradients. This effect causes poor performance in the winter months, especially when there is ice cover (Barrett, 2004b). The type and condition of vegetation within a BMP will determine the efficiency of phytoremediation mechanisms. In grassed swales and vegetated filter strips, filtration efficiency will also be affected by the type and condition of the vegetation. The climatological factors of temperature, solar radiation, season, and wind will effect the conditions within a BMP. Changes in temperature will cause changes in the kinetic rates of processes occurring within a BMP, such as adsorption/desorption, biological uptake, biological conversion, and natural degradation processes. Solar radiation on a quiescent permanent pool BMP will contribute to temperature pollution. Solar radiation will also increase pollutant removals through photolysis (Law and Band, 1998). Season will affect the vegetation within a BMP. During the growing season, nutrient uptake by vegetation will be at its peak, but during the fall die-back, vegetation will not uptake nutrients as readily and may even release some of the nutrients they had previously taken up. Released nutrients may be in the same form or a different form from what was originally taken up. Excessive wind can increase the dissolved oxygen in a pond or wetland type BMP. In shallow BMPs such as wet swales, wetlands, and shallow ponds, wind will also interfere with settling and may even resuspend previously settled solids. Possible groundwater intrusion into a BMP can contribute additional dissolved pollutant loadings or lead to dilution which can mask measured efficiencies when groundwater is not accounted for. Finally, BMP age and regularity of maintenance contribute variations in BMP pollutant removal efficiency. If not removed in a timely manner, decay of gross solids may contribute to pollutant loadings. Also, accumulation of sediments within a BMP will decrease efficiency by creating excessive bypassing or short circuiting in filters and pond-type BMPs, respectively.

5.1.4.2 Approximate Pollutant Removals of Structural BMPs

The pollutant removal of structural BMPs has been the subject of many studies. As stated earlier, the majority of reported data is presented in percent removals. This method of evaluating BMPs is widely criticized (see Section 4.5.4 for more information), however, a proven better measure for evaluating BMP pollutant removal efficiencies is not available and/or widely accepted (Urbonas, 2000; GeoSyntec and ASCE, 2002; U.S. EPA, 2002a; Clar *et al.*, 2003). Recently some have suggested that observed effluent quality is a more robust method for characterizing BMP performance (Strecker *et al.*, 2004). Therefore, the approximate pollutant removals of the most common structural BMPs are presented in terms of percent removals in Table 5-5 and observed effluent quality in Table 5-6 for a select group of pollutants. Prior to the presentation of Tables 5-5 and 5-6, it must be stressed again that due to the variable nature of BMPs, numerically reported efficiencies should be considered general estimates only.

Table 5-5. Approximate Pollutant Removal Efficiencies for Common Structural BMPs[1]

BMP Type	Percent Removal of Select Parameters				
	TSS	TP[2]	TN[2]	NO_x[2]	Metals
Dry extended-detention pond	61	19	31	9	26 - 54
Wet retention pond[3]	68 ± 10 (-33 - 99)	55 ± 7 (12 - 91)	32 ± 11 (-12 - 85)	34 ± 21 (-85 - 97)	36 - 65 (-97 - 96)
Infiltration trench	75	60 - 70	55 - 60	--	85 - 90
Porous pavement	82 - 95	65	80 - 85	--	98 - 99
Bioretention	80[4]	65 - 87	49	15 - 16	43 - 97
Sand or organic media filter	66 - 95	4 - 51	44 - 47	-95 - 22	34 - 88
Stormwater wetland[5]	71 ± 35	56 ± 35	19 ± 29	40 ± 68	0 - 57
Grassed swale[3]	38 ± 31 (-100 - 99)	14 ± 23 (-100 - 99)	14 ± 41 (-100 - 99)	13 ± 31 (-100 - 99)	9 - 62 (-100 - 99)
Vegetated filter strip[6]	54 - 84	-25 - 40	20[4]	-27 - 20	-16 - 55

[1]all data from (U.S. EPA, 2003b) unless otherwise noted
[2]TP = total phophorus; TN = total nitrogen; NO_x = nitrite + nitrate nitrogen
[3]summary of 35 studies for wet ponds and 20 studies for swales presented individually (U.S. EPA, 2003b): mean ± 95% confidence level (minimum - maximum)
[4](ARC, 2001)
[5]pond/wetland system
[6]lower number corresponds to 75 ft buffer; upper number corresponds to a 150 ft buffer

Although the data in Table 5-5 was taken primarily from one source (i.e., U.S. EPA, 2003b), the variability in percent removals is quite clear. This is especially true for wet retention ponds and grassed swales. The U.S. EPA's menu of BMPs (U.S. EPA, 2003b) listed numerous individual studies for these two categories. The variability in reported percent removals was extreme. In combining the listed data, the approximate percent removals had quite large confidence levels. The data in Table 5-6 is also taken from the U.S. EPA's menu of BMPs (U.S. EPA, 2003b) and although it appears to be less variable, a large degree of uncertainty still exists in many instances. Thus, the data in Tables 5-5 and 5-6 corroborate the point that caution is strongly advised when applying collected pollutant removal efficiency data to a different site.

Table 5-6. Observed Effluent Quality Performance for Common Structural BMPs[1]

BMP Type	Effluent Quality of Select Parameters							
	TSS (mg/L)	TP (mg/L as P)	TN (mg/L)	NO_x (mg/L as N)	Total Cu (µg/L)	Total Pb (µg/L)	Total Ni (µg/L)	Total Zn (µg/L)
Detention pond	32 ± 12 (2.5 - 140)	0.32 ± 0.04 (0.02 - 0.86)	1.8 ± 0.88 (0.45 - 6.0)	0.28 ± 0.05[2] (0.25 - 0.60)	20 ± 2.6 (0.5 - 82)	27 ± 6.4 (1.3 - 200)	4.8 ± 0.82 (0.5 - 11)	110 ± 18 (0.70 - 610)
Retention pond	24 ± 3.9 (0.03 - 250)	0.39 ± 0.15 (0.01 - 22)	1.1 ± 0.09 (0.10 - 4.1)	0.12 ± 0.03 (0.00 - 2.0)	11 ± 1.7 (0.13 - 130)	14 ± 2.5 (0.13 - 130)	4.0 ± 0.96 (0.11 - 23)	24 ± 2.9 (1 - 350)
Infiltration trench	240 ± 260[2] (120 - 420)	NI[3]	2.0 ± 0.76[2] (1.4 - 2.3)	0.57 ± 0.32[2,4] (0.36 - 0.73)	NI	26 ± 25[2] (12 - 42)	NI	90 ± 84[2] (50 - 150)
Porous pavement	24 ± 20[2] (0.55 - 52)	0.03 ± 0.01[2] (0.02 - 0.04)	2.9[5]	0.55[4,5]	8.4 ± 17[2] (0.24 - 30)	14 ± 12[2] (0.91 - 33)	NI	11 ± 8.8[2] (1.5 - 26)
Sand/Media filter	11 ± 7.4 (2.5 - 55)	0.24 ± 0.04 (0.002 - 2.3)	NI	0.26 ± 0.01[2] (0.25 - 0.34)	17 ± 2.7 (1.2 - 150)	8.4 ± 1.6 (1.0 - 110)	4.5 ± 8.7 (2.0 - 22)	120 ± 22 (1.0 - 960)
Wetland Basin	22 ± 8.1 (0.14 - 730)	0.13 ± 0.02 (0.00 - 1.4)	2.0 ± 0.60 (0.01 - 51)	0.22 ± 0.10 (0.00 - 8.1)	3.9 ± 0.53 (0.50 - 16)	1.6 ± 0.29 (0.10 - 10)	NI	47 ± 13 (2.0 - 500)
Grassed swale	26 ± 23 (5.0 - 56)	0.28 ± 0.08 (0.04 - 2.7)	0.62 ± 0.10 (0.06 - 2.7)	0.85 ± 0.86 (0.08 - 3.7)	6.5 ± 1.3 (0.30 - 56)	4.3 ± 1.1 (0.50 - 33)	NI	34 ± 3.7 (4.0 - 150)
Vegetated filter strip	37 ± 47[2] (5.0 - 56)	0.91 ± 0.62 (0.15 - 9.3)	NI	NI	7.2 ± 1.7 (1.0 - 17)	15 ± 12 (1.0 - 150)	2.4 ± 0.76[2] (2.0 - 4.3)	44 ± 11 (3.0 - 150)

[1] all data from the National Stormwater BMP Database (http://www.bmpdatabase.org/): mean ± 95% confidence level (minimum - maximum)
[2] based on one study only
[3] No Information
[4] nitrate only
[5] one data point only

5.1.4.3 Approximate Pollutant Removals of Nonstructural BMPs

The nonstructural BMPs listed in Table 5-1, under Minimum Control Measure #5, will provide pollutant removal through source reduction or pollution prevention measures. Public education and participation measures foster better personal decisions on issues that affect stormwater runoff quality. The success of source controls that depend on public buy-in such as pet waste management, littering, lawn care and landscaping practices, vehicle washing practices, and used oil and hazardous waste collection and recycling programs rely heavily on education and participation of the community. Effectiveness of these essential, but indirect BMPs is difficult to assess. Based on self-reported changes in behavior, intensive participation programs will be more effective at reducing pollutant loadings, than passive education measures such as brochures (Taylor and Wong, 2002a). Further, more memorable educational measures such as theater productions about stormwater pollution issues and creative signage will be more likely to affect change.

An additional nonstructural BMP of an effective stormwater management program is the detection and elimination of illicit connections to stormwater drainage systems. Illicit connections may include improper sanitary sewage connections, effluent from improperly operating septic tanks, leaking tanks or pipes from industrial sites and wastewater from commercial car wash and laundry facilities (Pitt *et al.*, 1993). When present, these discharges may result in significant pollutant loadings. For example, the Wayne County Department of Public Health estimated that their illicit connection elimination program kept over 2,000 pounds of suspended solids from entering the storm sewer system over a four year period (U.S. EPA, 2002b). Depending on the type of illicit connections identified, other pollutants that may be reduced by eliminating these connections include oxygen demanding substances, bacteria and other pathogens, hydrocarbons, and toxic chemicals. Since the existence and relative volume of illicit connections are highly site specific, the effectiveness of this BMP is also highly site specific. Pitt *et al.* (1993) and U.S. EPA (1999c) discuss methods of detecting illicit discharges and appropriate corrective measures.

Construction site runoff management using erosion and sediment controls have been reported to be very effective at preventing solids from leaving the construction site. The average suspended solids load reduction by erosion controls such as preventative construction planning and phasing and intensive site mulching and seeding has been estimated as approximately 85%, while sediment controls such as sediment fences and sediment basins are reputed to achieve approximately 60 to 70% (Taylor and Wong, 2002a).

Although, the pollution removal efficiencies of most pollution prevention controls cannot be measured directly, some studies have produced estimates of the water quality control benefits of these nonstructural BMPs. A summary of this information is presented in Table 5-7. Also, as mentioned above, a number of these measures are dependent on personal behaviors, thus the effectiveness of public education and outreach and public involvement and participation initiatives will indirectly affect the actual realized benefits.

Table 5-7. Generalized Water Quality Performance of Select Nonstructural BMPs[1]

BMP Type	Qualitative Control Information	Estimated Percent Load Reduction
Pet waste collection	potential load reductions for bacteria	none available
Vehicle washing	untreated car wash effluent is of poor quality (mean concentrations in mg/L from one study were copper = 0.386, lead = 0.113, total nitrogen = 4.11, total phosphorus = 0.32, TSS = 178, and zinc = 0.387); loading of these pollutants may be reduced if vehicle washing is performed on a vegetated area where effluents have an opportunity to infiltrate; removal could approach 100% if local ordinances require washing at facilities where effluents are drained to sanitary sewer lines	none available
Landscaping and lawn care	choosing landscaping plants and designs with low maintenance regimes can decrease runoff volume through reduced watering requirements and reductions in nutrient loadings through reduced fertilization requirements; lawn and garden fertilization is estimated to contribute 80 - 85% of phosphorus loading in the spring	none available
Storm drain system cleaning	removal of grass clippings and leaves from inlet structures can reduce organic carbon and nutrient loadings; significance of impact will depend on the land use of the drainage area	copper: 3 - 4%
Roadway and bridge maintenance	preventative maintenance and prompt repairs of structural degradation will reduce solids loadings to runoff	none available
Pest control	integrated pest management techniques can reduce pesticide and herbicide usage by 75 - 87% and 85 - 90%, respectively; information on how this translates to reduction in runoff loads is not available	none available

Table 5-7. Generalized Water Quality Performance of Select Nonstructural BMPs[1]

BMP Type	Qualitative Control Information	Estimated Percent Load Reduction
Parking lot and street cleaning	impact on organic carbon and nutrient loadings may be high during fall season due to deciduous leaf removal; access to the curb must be maintained using enforced parking restrictions for maximum effect; actual efficiency is highly dependent on sweeping frequency; frequencies need to be approximately once a week to maintain non-negligible effect	cadmium, copper, and zinc: 45% chemical oxygen demand (COD): 34 - 45% iron: 13 - 60% lead: 5 - 48% litter: 95 - 100% organic nitrogen: 12 - 45% total phosphorus: 9 - 28% TSS: 37 - 50% soluble phosphorus: 45%
Industrial good housekeeping practices	improved materials handling and storage practices can reduce sediment and chemical loadings; the data shown are the estimated effects of improved industrial housekeeping measures (education of managers and implementation of an auditing program using a before/after study design), the estimated largest measure taken was moving one of three industrial material stockpiles in the catchment area from an uncovered to a covered area	copper: 42% lead: 72% total nitrogen: 40% total phosphorus: 49% TSS: 8% zinc: 83%
Catchbasin cleaning	removal of grass clippings and leaves from inlet structures can reduce organic carbon and nutrient loadings; significance of impact will depend on the land use of the drainage area	COD, total phosphorus, total kjeldahl nitrogen (TKN), and zinc: 5 - 10% lead and TSS: 10 - 25%
Used oil and hazardous chemicals collection/recycling	assuming that all household hazardous wastes and used oil would otherwise be dumped down storm drains, tens to hundreds of tons of these materials could be prevented from causing receiving water pollution	none available

[1]Source of data: (Taylor and Wong, 2002a)

5.1.5 Maintenance, Reliability, and Cost Issues

Costs, including capital, operation and maintenance, and time and effort requirements associated with maintenance can be determining factors in the ultimate selection of an appropriate BMP. Specific costing issues are covered in detail in Chapter 6 of this document and, therefore, will not be repeated here. Thus, this section will focus on BMP reliability and maintenance regimes necessary to sustain BMP effectiveness.

One strong factor in BMP selection is reliability of the BMP technology. Planning and installation of a BMP can be costly and time consuming. To achieve the expected benefits and avoid wasted efforts, the chosen BMP must be reasonably reliable. Levels and factors relating to the reliability of the available BMP technologies are presented in Table 5-8.

The level of maintenance can vary significantly from one BMP to the next. Factors that affect the required level of maintenance include, BMP type, visibility of the BMP, landscaping, upstream conditions and safety issues (NJDA *et al.*, 2000). If not maintained properly, the reasons for selection of the particular BMP, such as it's basic level of pollutant removal performance and flood protection, will be lost. For instance, if sediment is allowed to accumulate in the bottom of a retention pond, eventually the hydraulics within the pond will change and short-circuiting and/or a loss in treatment capacity may occur, thereby reducing the pollutant removal efficiency (Law and Band, 1998). The accumulated sediment also will decrease the storage volume of the pond, which in turn will decrease the flood control value of the pond.

Operational needs of BMPs include visual inspection, which should be done annually as a minimum and after intense storms (NJDA *et al.*, 2000). Inspection checklists will be BMP- and site-specific. Some maintenance activities will be driven by these inspections. For example, bank stabilization of emergency spillways should only occur if inspectors find that the bank is beginning to show signs of erosion. Other examples include cleanup of vandalism, removal of burrowing animals, and seeding or sodding in areas of dead or damaged ground cover. These activities are usually unplanned and sporadic in nature and as such are not considered selection related. Other maintenance activities are expected occurrences that require routine attention and thus can be factored into the overall cost of a BMP (as seen in Chapter 6). These maintenance activities are described for some commonly used BMPs in Table 5-8. In summary, a higher maintenance burden can be expected with infiltration basins, infiltration trenches, porous pavement, and stormwater wetlands. A medium maintenance burden can be expected with dry extended retention ponds, wet ponds, sand and organic media filters, catchbasins and catchbasin inserts, manufactured products for stormwater inlets, and modular treatment devices. A low maintenance burden can be expected with bioretention, grassed swales, vegetated filter strips, greenroofs, rain gardens, dry wells, and rain barrels/cisterns. Finally, in-line storage devices carry little maintenance burden at all since they are designed to be self-cleaning. However, they should be inspected regularly for malfunction and/or clogging.

Nonstructural BMPs (with a few exceptions, such as street and catchbasin cleaning) are not generally thought of as having maintenance issues in the usual sense of the word, however, they do require ongoing effort. Education and public participation programs must continually evolve to remain productive and reach new audiences. Although genuine participation programs are more effective than purely educational programs, they require a much greater time commitment. Participation programs require the fostering of partnerships between agencies and community stakeholder groups (Taylor and Wong, 2002a). To maintain these partnerships in proper working order takes continuing time and effort.

Table 5-8. Generalized Structural BMP Reliability and Maintenance Requirements[1]

BMP Type	Reliability	Required Maintenance Activities	Recommended Maintenance Interval	Consequences of Failing to Perform Maintenance
Dry extended-detention pond	moderate to high; design criteria well established, but wide range in performance indicates room for improvement	1. mow side slopes and remove litter	monthly	degradation of aesthetics (actual frequency will depend on climate)
		2. remove debris from inlet and outlet structures and repair damage if necessary	monthly and/or following large storm events	inlet blockage - possible upstream flooding; outlet blockage - extended ponding within BMP which may encourage mosquito breeding
		3. repair and re-vegetate eroded area	annually (if required)	loss of treatment value and increase in effluent sediment loading
		4. remove sediment from the forebay	5 - 7 yrs	decrease in maintenance interval for the more energy and cost intensive maintenance activity #5
		5. remove sediment across the entire pond bottom	25 - 50 yrs	increased resuspension such that pond becomes pollutant source and decrease in flood control volume

Table 5-8. Generalized Structural BMP Reliability and Maintenance Requirements[1]

BMP Type	Reliability	Required Maintenance Activities	Recommended Maintenance Interval	Consequences of Failing to Perform Maintenance
Wet retention pond	moderate to high; design criteria well established, but wide range in performance indicates room for improvement	1. mow side slopes and remove litter and/or hydrocarbon build-up	monthly	degradation of aesthetics; hydrocarbon build-up may damage plants and pond microbiology (actual frequency will depend on climate and contributing land use)
		2. remove debris from inlet and outlet structures	monthly	inlet blockage - possible upstream flooding; outlet blockage - local flooding
		3. manage and harvest wetland plants (if included in design) and remove invasive species	annually (if required)	may be required in nutrient sensitive areas where seasonal senescence can lead to significant annual nutrient releases; invasive species may decrease wildlife habitat
		4. remove sediment from the forebay (after ~50% loss of total capacity)	5 - 7 yrs	decrease in maintenance interval for the more energy and cost intensive maintenance activity #5
		5. remove sediment across the entire pond bottom (after ~25% loss of permanent pool volume)	20 - 50 yrs	increased resuspension such that pond becomes pollutant source and decrease in flood control volume
		6. repair inlet/outlet structures and undercut or eroded areas	infrequently (as needed)	loss of treatment value; possible upstream of localized flooding

Table 5-8. Generalized Structural BMP Reliability and Maintenance Requirements[1]

BMP Type	Reliability	Required Maintenance Activities	Recommended Maintenance Interval	Consequences of Failing to Perform Maintenance
Infiltration trench	low to moderate; reported failure rates of about 50%, actual failure rates estimated to be higher	1. rake out and remove any accumulated sediment and debris from trench and inlets	monthly	degradation of aesthetics; infiltration will be impeded, leading to loss of treatment value and possible flooding
		2. replace top layers of pea gravel, filter fabric, and topsoil	as needed (when clogged)	caking of top layers leading to excessive bypass
		3. total rehabilitation to maintain a minimum of 2/3rds the design storage capacity and a 72 hr exfiltration rate	upon failure (typically within 5 yrs)	decrease in infiltration rates leading to loss of treatment value, extended ponding which may encourage mosquito breeding, and possible flooding
		4. excavate trench walls to expose clean soil	upon failure	decrease in infiltration rates

Table 5-8. Generalized Structural BMP Reliability and Maintenance Requirements[1]

BMP Type	Reliability	Required Maintenance Activities	Recommended Maintenance Interval	Consequences of Failing to Perform Maintenance
Porous pavement	low to moderate (modular type); low (monolithic type) which clogs much more rapidly	1. vacuum sweep (and high pressure wash monolithic) the surface, removing any obstructing sediment and debris	as needed (typically quarterly)	decrease in infiltration rates leading to loss of treatment value and possible flooding
		2. stabilize adjacent areas	annually (if required)	unstable adjacent areas will contribute excess sediment loading, reducing effective life of pavement
		3. replace deteriorating or spalling modular pavement	annually (if required)	breakdown of pavement will contribute to clogging and premature deterioration of adjacent fully operational pavement
		4. relieve localized clogging by drilling ½ in. holes through pavement every few feet within affected spot	upon localized failure	loss of treatment value; localized flooding; decrease in maintenance interval for the more energy and cost intensive maintenance activity #5
		5. total replacement	upon complete failure (1-2 yrs for monolithic)	complete loss of treatment value and possible flooding

Table 5-8. Generalized Structural BMP Reliability and Maintenance Requirements[1]

BMP Type	Reliability	Required Maintenance Activities	Recommended Maintenance Interval	Consequences of Failing to Perform Maintenance
Bioretention	no information	1. remove any litter and debris	monthly	degradation of aesthetics
		2. remove and replace dead or diseased vegetation	biannually	degradation of aesthetics, loss of treatment value
		3. add fresh mulch	annually	increased erosion due to exposed soil
		4. remove and replace entire mulch layer	2 - 3 yrs	loss of treatment value due to breakdown of mulch material and decrease in ion-exchange capacity
		5. repair of structural features (e.g., underdrains)	infrequent (as needed)	localized flooding; lack of drainage may kill plant root systems

Table 5-8. Generalized Structural BMP Reliability and Maintenance Requirements[1]

BMP Type	Reliability	Required Maintenance Activities	Recommended Maintenance Interval	Consequences of Failing to Perform Maintenance
Sand or organic media filter	low to moderate; organic media filters such as peat, peat/sand mix, and compost/ sand mix tend to clog faster than sand alone	1. remove debris/floatables from inlet/outlet structures and sedimentation chamber (if present)	monthly	inlet blockage - possible upstream flooding; outlet blockage - extended ponding within BMP which may encourage mosquito breeding; loss of treatment value if filter is bypassed due to blockage at either inlet or outlet
		2. remove sediment from sedimentation chamber (if present)	as needed (when 50% full)	inadequate sedimentation will lead to accelerated filter clogging
		3. renew or replace top few inches of filter bed surface	annually (or as needed)	filter bed sealing; excessive bypass volumes; loss of treatment value
		4. renew or replace entire filter bed (once draw down exceeds 24 hrs)	3 - 5 yrs	filter bed clogging; excessive bypass volumes; loss of treatment value; and extended ponding
		5. repair leaks or deterioration of structural components	infrequent (as needed)	loss of treatment value; possible localized flooding

Table 5-8. Generalized Structural BMP Reliability and Maintenance Requirements[1]

BMP Type	Reliability	Required Maintenance Activities	Recommended Maintenance Interval	Consequences of Failing to Perform Maintenance
Stormwater wetland	moderate to high; design criteria well established, but wide range in performance indicates room for improvement	1. mow side slopes and remove litter	monthly	degradation of aesthetics (actual frequency will depend on climate)
		2. remove debris from inlet and outlet structures	monthly	inlet blockage - possible upstream flooding; outlet blockage - localized flooding
		3. remove and replace dead or diseased vegetation	annually (if required)	loss of treatment value; degradation of aesthetics; coverage of ≥50% of wetland surface area is recommended
		4. harvest and replace wetland plants	annually (if required)	may be required in nutrient sensitive areas where seasonal senescence can lead to significant annual nutrient releases, however benefits of regular harvesting are not well substantiated by field data
		5. remove and replace invasive vegetation	annually (if required)	loss of wildlife habitat
		6. remove sediment from the forebay (after ~50% loss of total forebay capacity)	5 - 7 yrs	decrease in maintenance interval for the more energy and cost intensive maintenance activity #7
		7. remove sediment across the entire pond bottom (after ~25% loss of wetland storage volume)	20 - 50 yrs	decrease in flood control volume, increase in resuspension potential such that wetland may become pollutant source, vegetation may be choked out, and wetland may become overly eutrophic leading to aesthetic and odor problems
		8. repair inlet/outlet structures and undercut or eroded areas	infrequently (as needed)	loss of treatment value; possible upstream of localized flooding

Table 5-8. Generalized Structural BMP Reliability and Maintenance Requirements[1]

BMP Type	Reliability	Required Maintenance Activities	Recommended Maintenance Interval	Consequences of Failing to Perform Maintenance
Grassed swale	moderate; subject to disruption by homeowners	1. mow side slopes and remove litter and grass clippings	monthly	degradation of aesthetics (actual frequency will depend on climate); maintain height at 3 - 6 in. to retain filtration value
		2. remove and replace dead or diseased vegetation	annually (if required)	loss of treatment value; degradation of aesthetics
		3. aerate and dethatch grass on swale bottom	annually (once draw down takes > 48 hrs)	decrease in infiltration rates leading to loss of treatment value, extended ponding which may encourage mosquito breeding, and possible flooding
		4. remove sediment accumulation from swale bottom once treatment volume reaches 25% of design volume	5 yrs (or as needed)	decrease in flood control volume and decrease in infiltration rates
Vegetated filter strip	moderate; concentration of flows can often lead to erosion and failure	1. mow grass and remove litter and grass clippings	monthly	degradation of aesthetics (actual frequency will depend on climate); maintain height at 2 - 4 in. to retain filtration value
		2. remove and replace dead or diseased vegetation	annually (if needed)	loss of treatment value; degradation of aesthetics
		3. repair (reseed or resod) any rills, gullies, or bare areas	annually (if needed)	loss of treatment value due to short-circuiting
		4. rototill or cultivate surface of sand/soil bed	infrequent (as needed)	decrease in infiltration rates leading to loss of treatment value

[1]Source of data: (NJDA *et al.*, 2000; Urbonas, 2000; ARC, 2001; NYSDEC, 2001; U.S. EPA, 2002a; U.S. EPA, 2003b)

5.1.6 Environmental and Community Acceptance Factors

The final group of factors to consider before definitive selection of a BMP is environmental and community acceptance factors. Environmental factors include the possibility of mosquito breeding and providing wildlife habitat. Detention (dry) ponds, swales, stormwater wetlands, and infiltration BMPs have the potential to become mosquito breeding grounds, especially if they are not properly maintained. Before BMP installation, consider performing an assessment of the local wildlife and their habitat needs. This may be more important in areas of new development where the impact on available habitat is more pronounced. Natural, unmowed vegetation found in stormwater wetlands, some swales, and vegetated wet ponds, can provide habitat for song birds and small animals, whereas cleared and manicured BMPs, like detention (dry) ponds and infiltration basins provide no wildlife habitat value (NJDA *et al.*, 2000).

The two key community acceptance factors are: (1) safety issues; and (2) aesthetics, especially as they relate to financial implications. Community acceptance is especially important if public participation is required. Safety issues with BMPs are rare and those that do exist can usually be resolved through good design. An example of this is installation of wire grates on outlet structures to prevent sudden storm surges from sweeping someone into a storm drain. Another safety concern is the drowning hazard posed by permanent volume BMPs (wet ponds and wetlands). This can be reconciled by installing a fence around the perimeter of the BMP, however, this will make maintenance more difficult and also detract from the aesthetic value of these BMPs. Thus, community acceptance of permanent volume BMPs may be difficult to attain in areas with a high density of young children.

Aesthetic issues are of particular importance for community acceptance, not only because the property owners have to live side-by-side with the BMP, but also due to the financial implications aesthetic value brings. BMPs that are aesthetically pleasing will appreciate the value of the surrounding properties, and vice-versa. Thus aesthetics are often a key factor in gaining community acceptance. Although aesthetics are usually a matter of personal preference, a few aesthetic issues are more generally accepted. Odor nuisances caused by failing infiltration type BMPs and over eutrophication of wet ponds and wetlands can be problematic. Cattails and other emergent wetland vegetation in wet ponds, wetlands, and some swales are considered unsightly to some people (NJDA *et al.*, 2000) while not to many others. BMPs such as infiltration basins and trenches and rain gardens, which are intended to create ponding areas that drain over a relatively short period of time, can be either objectionable "soggy spots" or desirable scenery depending on the viewer. Dry detention basins, especially if not well maintained, often become large garbage collectors. On the other hand, BMPs like wet ponds, stormwater wetlands, bioretention areas, rain gardens, and greenroofs can give a community an up-scale look, adding to the aesthetic value of a property rather than detracting from it.

Public acceptance is paramount for BMPs that require public participation, such as rain barrels and dry wells, or nonstructural BMPs like car washing and pet waste management ordinances. Studies have shown that "the magnitude of induced behavioral change is likely to vary greatly on the nature of the activity promoted and the target audience" (Taylor and Wong, 2002a). As an example, Taylor and Wong (2002a) contend that an educational campaign to get people to wash their cars on their lawns would have a greater level of public acceptance in a low-density residential area than one that encourages the use of sewered wash bays. If one can attach a personal benefit to a BMP that requires public participation, then public acceptance will be higher and adoption of the BMP will occur at a much greater rate (Taylor and Wong, 2002a). For instance, playing up the costs savings on water bills gained from using a rain barrel for landscape watering needs will improve public acceptance of this BMP.

5.2 Structural BMP Placement

BMP placement refers to the spatial scale covered by a single BMP or set of BMPs used to protect the watershed as a whole. Three spatial scales of BMP placement have been identified; onsite, sub-regional, and regional (Lai *et al.*, 2003). The onsite spatial scale is defined as 10 to 100 acres, a size typical of a single commercial/industrial building lot or residential neighborhood. The sub-regional scale is defined as 100 to 5,000 acres and represents the township level. The regional scale has a drainage area greater than 5,000 acres, exemplifying an entire watershed, sewershed, or county. There is currently no comprehensive way to make the decision between a collection of small onsite BMPs located throughout the watershed or one large regional control located downstream or a combination of the two, although research in this area is underway (Sullivan *et al.*, 2003). The integrated stormwater management decision-support framework, discussed earlier in Section 5.1, is to include BMP placement as a central part of its overall BMP selection tool. Issues such as retrofit versus new development, cost, land availability, and hydrologic/hydraulic/water quality routing principles will be factored into BMP placement decisions (Lai *et al.*, 2003). However, this project is still in its initial phase. In the meantime, some general guidelines on BMP placement are available.

The issue of BMP placement is largely affected by whether the watershed in question is an existing urban area or a new development. Retrofitting BMPs into an existing urban area places many constraints on BMP placement. Existing drainage systems, limited available space, and the high cost of land in these areas make applying large, regional BMPs much more difficult in existing developments (Sieker and Klein, 1998). However, in some circumstances these facilities have been shown to be cost effective (Strecker *et al.*, 2002). Densely populated areas are the most complicated to retrofit. The high percent imperviousness of such areas requires rapid conveyance of the stormwater in order to avoid flooding (U.S. EPA, 1999a). Thus, the question of BMP placement within high density areas is limited to source control techniques, such as greenroofs, in-line storage management techniques in existing conveyance systems, or retrofit of existing flood conveyance or detention basins. In existing lower density areas, some onsite BMPs become feasible, such as use of rain barrels and dry wells. The expensive retrofit of replacing curb and gutter systems with swales can also be used in existing low density areas. Sand and media filtration systems are one regional control option that is often used in retrofit situations due to its limited space requirements (U.S. EPA, 1999a). Sand filtration can even be constructed underground thus requiring a very minimal footprint. However, larger BMPs traditionally used as regional controls, such as dry and wet ponds, are usually not an option. Despite these constraints, areas as densely populated as New

York City's Staten Island have instituted a wide-reaching BMP program that successfully addressed placement issues (Vokral *et al.*, 2001). The New York City Department of Environmental Protection (NYCDEP) developed the Staten Island Bluebelt Program to combat flooding problems. A knowledge of flood prone areas was combined with a review of the regional topography, the locations of existing storm sewers, and the ecological condition and hydrological patterns of local streams. This assessment assisted the NYCDEP in the placement of BMP retrofits. Once high risk areas were identified, property in the area was acquired and a system of small "pocket wetlands" were installed in areas as small as single lots.

In new developments, the use of BMPs are more of a proactive approach rather than a reactive approach. Thus, the issue of proper placement of BMPs in the watershed is more of a conscious decision rather than a reaction to existing consequences as with retrofit situations. Onsite practices can be advantageous because they combat the problem close to its source. This makes onsite placement a good option for approximating pre-development hydrology. The onsite scale is most often used due to its ease of implementation (Clar *et al.*, 2003). Due to the small area involved, onsite placement can reduce political or institutional complications. Many municipalities have regulations requiring the use of onsite BMPs as part of their permits for new construction. This has the benefit of shifting financial responsibility to the developer. Yet, onsite placement is not without its problems. The decentralized ownership of onsite BMPs leads to a wide range of BMP design, construction, and operation and maintenance practice issues (Finnemore, 1982). In addition, options to treat existing runoff together with that from new development could be missed. Under these circumstances, water quality and quantity control benefits are difficult to assess and may be unreliable. Also, onsite placement of BMPs is sometimes difficult to coordinate in order to reap real stream protection benefits on a watershed scale.

Sub-regional and regional BMP placement options can be advantageous for several reasons (Clar *et al.*, 2003). Their affect on the receiving water is more easily observed and quantified. For watershed management programs under the control of a public agency, sub-regional or regional placement can be more cost-effective because one large BMP will cost less to build and maintain than many smaller BMPs. Maintenance programs for BMPs at this level are more likely to be consistent and long-term. And finally, the size of regional BMPs is conducive for recreational, aesthetic, and wildlife habitat benefits that may not necessarily accompany smaller onsite BMPs. Disadvantages to the wider scales of BMP placement include: (1) the need for comprehensive; advanced planning; (2) large initial capital costs; (3) failure to provide significant flood relief; (4) failure to provide downstream channel protection through peak discharge control; (5) blockage of fish migration routes; (6) possible thermal pollution; and (7) accelerated erosion and water quality degradation in the unprotected upstream feeder streams (Urbonas, 2000; Clar *et al.*, 2003).

5.3 INTEGRATED APPROACH

An integrated approach to stormwater management is considered more effective than any single BMP alone. Case studies recognize that in-stream biological condition may not be measurably protected or improved by an isolated BMP (Stribling *et al.*, 2001). Stream degradation is rarely caused by a single stressor and more than one type of BMP may be necessary to relieve all contributing stressors. The NPDES Stormwater Phase I and II Regulations mandate an integrated approach by requiring elements from each of the six minimum control measures shown in Table 5-1. Thus, evaluating the performance of an individual BMP may misrepresent the role of that BMP in the overall management scheme (Taylor and Wong, 2002a). Integration of BMPs may occur on three different levels: integration of more than one structural BMP, integration of more than one nonstructural BMP, and finally integration of structural and nonstructural BMPs. The following sections review each of these three levels.

The integrated approach discussed in this section is the integration of individual BMPs with each other. However, a holistic integrated approach also should consider BMP integration within the community. For example, incorporating small design features into the larger structural BMPs can turn them into public amenities (Roesner, 1999). Parks and walking paths around stormwater management facilities can add aesthetic and recreational value, turning BMPs from community nuisances into community assets.

Finally, it is important to acknowledge that measured pollutant removal efficiencies (specifically, percent removals) of individual BMPs may not be representative of the overall benefit to the receiving water when the BMP is part of an integrated approach (Rushton, 2002). When stormwater runoff is pre-treated by upstream BMPs, the influent water quality will be cleaner and thus a large difference between influent and effluent pollutant concentration will become more difficult to achieve. Calculating percent removals in these situations will produce deceiving results and lead to unwarranted perceptions about the worth of the BMP. Influent and effluent concentrations should always be reported in these cases. Rushton (2002) suggests comparing the BMP effluent concentrations to applicable water quality standards. Still, even when the BMP is part of a integrated system, monitoring of individual BMPs may be necessary to produce data for modeling. Watershed modeling often requires efficiency monitoring data for each BMP separately for the development of reliable statistical (black-box) models and/or the verification/calibration of process-based models.

5.3.1 Integration of Structural BMPs

Structural BMPs can be used together in an integrated approach in two distinct ways. Individual BMPs can be: (1) strung together in a directly connecting manner, i.e., the treatment train approach; or (2) operated separately, but used collectively in the protection of the same receiving water or watershed system.

Many possible treatment train options exist. Generally, smaller BMPs are used as pretreatment for and/or conveyance to larger BMPs. Using smaller BMPs as pretreatment has the added benefit of extending the maintenance intervals and useful life of the primary BMP. Examples of such BMP treatment trains are as follows:

- vegetated filter strips → sand filter or infiltration basin,
- grassed swales → dry extended-detention pond, wet pond, or wetland,
- bioretention → dry extended-detention pond or infiltration basins, and
- manufactured products for stormwater inlets or catchbasins and catchbasin inserts → dry extended-detention ponds or infiltration basins.

Using structural BMPs in treatment trains, such as those listed above, can improve both the water quality as well as total runoff volume of the final effluent, thereby minimizing the effects on receiving waters (U.S. EPA, 1999a). Unlike the unit processes in conventional drinking water or wastewater treatment trains that are engineered to perform a specific task, such as sedimentation or disinfection, individual structural BMPs are often not limited to one pollutant removal mechanism, as discussed in Chapter 2. However, the most effective combinations are made between BMPs with different dominant pollutant removal mechanisms, such as integration of a sand filter (filtration) with a wet retention pond (sedimentation). Combining strong water quantity control BMPs (e.g., dry detention ponds) with strong water quality control BMPs (e.g., sand filters) will also make effective treatment trains.

The multi-chambered treatment train (MCTT) is an example of an engineered BMP that integrates several pollutant removal mechanisms used by various BMPs (Greb *et al.*, 2000). This manufactured device is akin to a miniature wastewater treatment plant with three main chambers, a grit chamber, a settling chamber, and a media filter chamber. The system approximates a three component treatment train starting with a catchbasin, then a small dry extended-detention pond and polishing effluent with a media filter. The MCTT is designed to be installed underground and one study found appreciable pollutant removals of TSS, total phosphorus, and total recoverable phosphorus (Greb *et al.*, 2000). However, this treatment train will not provide significant water quantity control.

The Staten Island Bluebelt program discussed in Section 5.2 is an example of separate, but collectively used structural BMP integration. Their pocket wetlands are a combination of ponds and shallow marsh lands (Vokral *et al.*, 2001). These BMPs were engineered to act as a buffer between piped runoff and natural streams or drainage corridors. Individually each BMP provides some pollutant removal and peak runoff attenuation, but combined they provide a very effective system for stream protection. Also, since the individual parcels of land are often adjacent to each other, the Bluebelt BMP system provides a continuous corridor for a healthy wildlife habitat.

5.3.2 Integration of Nonstructural BMPs

Given the shear number of nonstructural BMP options, listing every possible combination of nonstructural BMPs would be impractical. Instead, four examples of nonstructural BMP integration are presented in the following paragraphs.

The benefit of the combined effect of sensitive construction planning and implementation of low impact development (LID) techniques can be seen in a comparison study produced by Finnemore (1982). The integrated set of BMPs used were open space requirements, cluster development of residential areas, strict avoidance of construction on areas with slopes steeper than 25%, and proper construction practices including, check dams and erosion baffles, lining drainage channels, filter fences and berms, and restoration of native vegetation. Suspended sediment loads were measured before and after residential development of two sites in Lake Tahoe, California. The well-planned, water sensitive site increased sediment loading by a factor of two and receiving water studies of macro-invertebrate communities detected a negligible impact, whereas the poorly planned site which implemented no BMPs increased sediment loadings by a factor of 107, causing an observed 34% decrease in macro-invertebrate density and a 54% decrease in the number of macro-invertebrate species. Thus, it was estimated that the benefit of this combination of construction site BMPs and LID measures was a 98% reduction in suspended sediment loading. The author of the study did caution that both the sites chosen for their study were extreme cases and that the impacts of most new development, whether BMPs are implemented or not, likely will fall somewhere in between the two they presented.

A municipality in Oklahoma instituted an integrated stormwater management program including the following BMPs; illegal discharge elimination program; litter collection campaign; illegal dumping minimization program; hazardous waste collection program; storm drain stenciling program; and an extensive advertizing campaign (Taylor and Wong, 2002a). Stormwater quality was monitored for four years both before and after implementation of the program. Collectively these BMPs produced average reductions in event mean concentrations of 13% TSS, 17% total phosphorus, 18% total kjeldahl nitrogen, and at least 55% for the heavy metals, copper, lead, and zinc. Although the data was not statistically validated, the results are quite promising.

When the effectiveness of education and outreach initiatives are viewed independently there is little evidence that "information alone can significantly change actual behavior" (Taylor and Wong, 2002a). However, use of an educational component as part of an integrated stormwater management program is integral to the success of many nonstructural BMPs. For example, ordinances will be ineffective unless preceded by an appropriate educational initiative (Taylor and Wong, 2002a). People cannot be expected to comply with a law if they don't know it exists. A review of over 100 case studies found that the effectiveness of other nonstructural BMP programs are often related to the effectiveness of education and outreach programs (Taylor and Wong, 2002a). The success of most integrated programs should improve if a well executed educational component is included. Yet, education and laws alone will not lead to compliance unless followed up by a strong enforcement program for incentive. The bigger picture is that ordinances by themselves are not a very effective BMP, but when integrated with education and enforcement, their effectiveness greatly improves.

5.3.3 Integrating Structural and Nonstructural BMPs

In general, structural BMPs are used much more readily than nonstructural BMPs. Structural BMPs are tangible entities that produce measurable results. They provide "the engineering calculations necessary to demonstrate compliance with numerical ... criteria" (NVPDC, 1996). However, most stormwater managers would agree that an effective stormwater management program requires an optimum balance between both structural and nonstructural BMP types (Taylor and Wong, 2002b). A comprehensive study of stormwater management programs concluded that effective integration of many types of BMPs produce the most successful results. Nonstructural practices that reduce the amount of runoff, such as cluster housing, reducing street width, and alternative turn-about designs, can reduce size requirements or numbers of structural BMPs. Nonstructural BMPs also can lessen pollutant loading thereby reducing the required maintenance burden of structural BMPs.

For these reasons, the concept of LID has gained a considerable amount of support. LID is often referred to as a single BMP, however, it is really an integrated concept containing both distributed structural and nonstructural components. Typical structural BMPs incorporated into LID include swales, vegetated filter strips, porous pavement, bioretention areas, and disconnection of roof leaders into rain barrels, dry wells, or rain gardens. Nonstructural BMPs commonly used in LID include construction site erosion and sediment controls, minimization of impervious areas, and pollution prevention and good housekeeping practices. Integration of these soft engineering distributed structural BMPs and nonstructural BMPs can result in a considerable reduction in pollutant loading.

Zielinski (2000a) completed a modeling study to compare LID to two conventionally designed residential developments with respect to percent imperviousness, stormwater runoff volume, and nutrient export. In a low-density residential area, LID elements (called open space design in the reference) included smaller lot size, narrower streets, shared driveways, a looped turnabout with central bioretention cell instead of the conventional cul-de-sac bulb, natural area protections, grassed swales, and improved septic tank design and installation. Modeling efforts showed reductions in average annual total nitrogen and phosphorus loads of 46 and 50%, respectively, and a 35% decrease in total impervious area over the conventional design without BMPs. In a medium-density residential area the LID elements were similar, including smaller lot size, narrower streets, vegetated buffer strips along waterways, shorter driveways, fewer sidewalks, elimination of turnabouts by using a looped road design, grassed swales, and several bioretention areas. A small central detention pond was also included in the open space design. This combination of LID elements and the pond resulted in a 24% reduction in total impervious area for the site. Furthermore, a significant reduction in average annual total nitrogen and phosphorus loads of 45 and 60%, respectively, was estimated for the open space design, even though the current conventional design included a large dry extended-detention pond.

In a related study Zielinski (2000b) explored LID practices for green parking lot designs. In one scenario, the green parking lot design included the following elements: reduction in the number and size of parking spaces, reduction in the width of the aisles, larger landscaping islands, some planted with shade trees, others used as bioretention areas, grassed swales, porous pavement (grid pavers), and a single wet retention pond to handle larger storms only. The conventionally designed parking lot included two large wet ponds for stormwater treatment. Still, modeling results estimated average annual total nitrogen and phosphorus load reductions of 45 and 47%, respectively, and a 22% reduction in total impervious area over the conventional design without BMPs. The results from the three scenarios investigated by Zielinski (2000a; 2000b), show that through the integration of structural and nonstructural BMPs better protection of the watershed is achieved than when each element is used individually.

LID supporters have highlighted other advantages of this micro-scale technique in addition to water quality improvements. LID's lot-level strategies can reduce hydrologic stresses of development. Pre-development hydrology is replicated as closely as possible by managing the stormwater runoff where it starts, sometimes eliminating the need for large scale centralized BMPs such as retention ponds (Lai *et al.*, 2003). Groundwater recharge is maximized by LID practices (Clar *et al.*, 2003). Finally, many of the techniques used in LID will reduce development costs and associated maintenance costs, such as having sidewalks on only one side of the street. Yet, LID is in its infancy compared to other BMPs and therefore it is not a proven approach. Disadvantages of LID include conflicts with some areas' building codes, difficulties gaining public support for new, unproven techniques, difficulties in constructing and maintaining and highly distributed structural controls, and the perception by some that LID promotes urban sprawl (Clar *et al.*, 2003).

REFERENCES

Atlanta Regional Commission (ARC). (2001). *Georgia stormwater management manual, Volume 2: technical handbook (first edition)*. Atlanta, GA. www.georgiastormwater.com. August 2001.

Barraud, S., Y. Azzout, F.N. Cres, and B. Chocat. (1999). Selection aid of alternative techniques in urban storm drainage - proposition of an expert system. *Water Science and Technology* **39**(4): 241 - 248.

Barrett, M.E. (2004a). Performance comparison of structural stormwater BMPs. *Water Environment Research*: awaiting publication.

Barrett, M.E. (2004b). Retention pond performance: examples from the international stormwater BMP database. In: *Proceedings of the World Water and Environmental Resources Congress - Critical Transitions in Water and Environmental Resources Management*, June 27 - July 1, 2004, Salt Lake City, UT.

Clar, M., B.J. Barfield, and S. Yu. (2003). *Considerations in the design of treatment best management practices (BMPs) to improve water quality*. EPA 600/R-03/103. U.S. Environmental Protection Agency, Office of Research and Development, Cincinnati, OH. www.epa.gov/ORD/NRMRL/Pubs/600r03103/600r03103.htm October 2003.

Commonwealth of Virginia (COV). (1999). *Virginia stormwater management handbook, Volume I (first edition)*. Department of Conservation and Recreation, Division of Soil and Water Conservation, Richmond, VA. http://ww.dcr.state.va.us/sw/docs/swm/Volume_I.pdf. Accessed on March 24, 2004.

Florida Department of Environmental Protection (FDEP). (2003). *Nonpoint source management - urban stormwater program*. Nonpoint Source Management Program, Tallahassee, FL. http://www.dep.state.fl.us/water/nonpoint/urban1.htm. April 2003. Accessed on January 21, 2004.

Federal Highway Administration (FHWA). (2000). *Stormwater best management practices in an ultra-urban setting: selection and monitoring*. FHWA-EP-00-002. U.S. Department of

Transportation, Federal Highway Administration, Washington, DC. http://www.fhwa.dot.gov/environment/ultraurb/index.htm.

Finnemore, E.J. (1982). Stormwater pollution control: best management practices. *Journal of Environmental Engineering Division, Proceedings of the American Society of Civil Engineers* **108**(EE5): 706 - 721.

Frost, A. (2003). Can BMPs really work in Scotland?, http://www.sepa.org.uk/guidance/urban-drainage/bmps.htm. Scottish Environmental Protection Agency (SEPA), Scotland, UK.

GeoSyntec Consultants Urban Drainage and Flood Control District and Urban Water Resources Council of ASCE (GeoSyntec and ASCE). (2002). *Urban stormwater BMP performance monitoring: a guidance manual for meeting the national stormwater BMP database requirements*. EPA 821/B-02/001. U.S. Environmental Protection Agency, Office of Water, Washington, DC.

Greb, S.R., R.T. Bannerman, S.R. Corsi, and R.E. Pitt. (2000). Evaluation of the multichambered treatment train, a retrofit water-quality management practice. *Water Environment Research* **72**(2): 207 - 216.

Lai, F.-h., C.-Y. Fan, L. Shoemaker, and R. Field. (2003). Development of a decision-support framework for placement of BMPs in urban watersheds. In: *National TMDL Science and Policy 2003 Specialty Conference*, Chicago, IL. Water Environment Federation.

Law, N.L. and L.E. Band. (1998). Performance of urban stormwater best management practices, http://www.unc.edu/depts/geog/them/projects/BMP.html. University of Carolina - Chapel Hill, Department of Geology, Chapel Hill, NC.

New Jersey Department of Agriculture, New Jersey Department of Community Affairs, New Jersey Department of Environmental Protection, and New Jersey Department of Transportation (NJDA *et al.*). (2000). *DRAFT revised manual for New Jersey: best management practices for control of nonpoint source pollution from stormwater (Fifth draft)*. http://www.state.nj.us/dep/watershedmgt/bmpmanual.htm. May 3, 2000.

Commission, N.V.P.D. (NVPDC). (1996). *Nonstructural urban BMP handbook - a guide to nonpoint source pollution prevention and control through nonstructural measures*. Prepared for the Department of Conservation and Recreation, Division of Soil and Water Conservation, Annandale, Virginia. http://novaregion.org/NSBMP1.

New York State Department of Environmental Conservation (NYSDEC). (2001). *New York state stormwater management design manual (Draft)*. Division of Water, Albany, NY. http://www.dec.state.ny.us/website/dow/swmanual. October 2001.

Parikh, P. (2003). A history and overview of the clean water act and its application to stormwater runoff. *National Risk Management Research Laboratory Seminar Series*, November 6, 2003, Cincinnati, OH.

Pitt, R., M. Lalor, R. Field, D.D. Adrian, and D. Barbe. (1993). *Investigation of inappropriate pollutant entries into storm drainage systems*. EPA 600/R-92/238. U.S. Environmental Protection Agency, Office of Research and Development, Cincinnati, OH. January 1993.

Roesner, L.A. (1999). Urban runoff pollution summary thoughts - the state-of-practice today and for the 21st century. *Water Science and Technology* **39**(12): 353 - 360.

Rushton, B. (1998). Processes that affect stormwater pollution, http://www.stormwater-resources.com/library/001PPocesses.PDF. Stormwater News, Cocoa Beach, Florida.

Rushton, B. (2002). BMP monitoring: methods and evaluations, http://www.stormwater-resources.com/library.htm#BMPs (#147PL). Stormwater News, Cocoa Beach, Florida.

Schueler, T.R. (1987). *Controlling urban runoff: a practical manual for planning and designing urban BMPs*. Metropolitan Washington Council of Governments, Department of Environmental Programs, Washington, DC. July, 1987.

Selvakumar, A. and M. Borst. (2003). Effects of land use and season on microorganism concentrations in urban stormwater runoff. In: *National TMDL Science and Policy 2003 Specialty Conference*, Chicago, IL. Water Environment Federation.

Sieker, H. and M. Klein. (1998). Best management practices for stormwater-runoff with alternative methods in a large urban catchment in Berlin, Germany. *Water Science and Technology* **38**(10): 91 - 97.

Strecker, E., P. Mangarella, and N. Brandt. (2002). A natural treatment system master plan approach to meeting TMDLs: case study of San Diego Creek, Orange County, California. In: *National TMDL Science and Policy Specialty Conference*, November 13 - 16, 2002, Phoenix, AZ. Water Environment Federation.

Strecker, E.W., M.M. Quigley, B. Urbonas, and J. Jones. (2004). Analyses of the expanded EPA/ASCE international BMP database and potential implications for BMP design. In: *Proceedings of the World Water and Environmental Resources Congress - Critical Transitions in Water and Environmental Resources Management*, June 27 - July 1, 2004, Salt Lake City, UT.

Stribling, J.B., E.W. Leppo, J.D. Cummins, J. Galli, S. Meigs, L. Coffman, and M.-S. Cheng. (2001). Relating instream biological condition to BMPs in watershed. In: *Linking Stormwater BMP Designs and Performance to Receiving Water Impact Mitigation*, August 19 - 24, Snowmass Village, CO. American Society of Civil Engineers.

Sullivan, D., R. Field, and C.-Y. Fan. (2003). The complications of monitoring treatment BMPs. In: *National TMDL Science and Policy 2003 Specialty Conference*, Chicago, IL. Water Environment Federation.

Taylor, A. and T. Wong. (2002a). *Non-structural stormwater quality best management practices - a literature review of their use, value, cost, and evaluation*. EPA Victoria Technical Report 02/11. Cooperative Research Center for Catchment Hydrology, Victoria, Australia. http://www.catchment.crc.org.au/pdfs/Technical200213.pdf December 2002.

Taylor, A. and T. Wong. (2002b). *Non-structural stormwater quality best management practices - an overview of their use, value, cost, and evaluation*. EPA Victoria Technical Report 02/13. Cooperative Research Centre for Catchment Hydrology, Victoria, Australia. http://www.catchment.crc.org.au/pdfs/technical200211.pdf December 2002.

United States Environmental Protection Agency (U.S. EPA). (1999a). *Preliminary data summary of urban storm water best management practices*. EPA 821/R-99/012. Office of Water, Washington, DC. August 1999.

United States Environmental Protection Agency (U.S. EPA). (1999b). *Stormwater management fact sheet - non-storm water discharges to storm sewers*. EPA 832/F-99/022. Office of Water, Washington, DC. http://www.epa.gov/owm/mtb/nonstorm.pdf. September 1999. Accessed on January 26, 2004.

United States Environmental Protection Agency (U.S. EPA). (2000). *Storm water Phase II final rule: an overview*. EPA 833/F-00/001. Office of Water, Washington, DC. January 2000.

United States Environmental Protection Agency (U.S. EPA). (2002a). *National management measures to control nonpoint source pollution from urban areas (Draft)*. EPA 842/B-02/003. Office of Water, Office of Wetlands, Oceans, and Watersheds, Nonpoint Source Control Branch, Washington, DC. September 2002.

United States Environmental Protection Agency (U.S. EPA). (2002b). *Illicit discharge detection and elimination - wastewater connections to the storm drain system*. Office of Water, Office of Wastewater Management, Washington, DC. http://cfpub.epa.gov/npdes/stormwater/menuofbmps/illi_7.cfm. August 2002. Accessed on January 26, 2004.

United States Environmental Protection Agency (U.S. EPA). (2003a). *Post-construction storm water management in new development and redevelopment - dry extended detention pond*. Office of Water, Washington, DC. http://cfpub.epa.gov/npdes/stormwater.menuofbmps/post_26.cfm. Accessed on August 27, 2003.

United States Environmental Protection Agency (U.S. EPA). (2003b). *National menu of best management practices for storm water Phase II*. Office of Water, Washington, DC. http://cfpub.epa.gov/npdes/stormwater/menuofbmps/menu.cfm. Accessed on August 27, 2003.

United States Environmental Protection Agency (U.S. EPA). (2003c). *Phase of the NPDES storm water program*. http://cfpub.epa.gov/npdes/stormwater/swphases.cfm. Accessed on March 30, 2004.

Urbonas, B. (2000). Assessment of stormwater best management practice effectiveness (chapter 7). In: *Innovative urban wet-weather flow management systems*, Field, R., J.P. Heaney, and R. Pitt, (editors). Technomic Publishing Co., Inc., Lancaster, PA. Pages 255 - 300.

Vokral, J., D. Gumb, R.D. Smith, and S. Mehrotra. (2001). Staten Island bluebelt program - a natural solution to environmental problems. *Stormwater* **2**(3): 58 - 69.

Wayland, R.H. and J.A. Hanlon. (2002). Memorandum: establishing total maximum daily load (TMDL) wasteload allocations (WLAs) for storm water sources and NPDES permit requirements based on those WLAs, http://www.epa.gov/npdes/pubs/final-wwtmdl.pdf. U.S. Environmental Protection Agency, Office of Water, Washington, DC. Accessed on August 18, 2004.

Water Environment Federation and American Society of Civil Engineers (WEF and ASCE). (1998). *Urban runoff quality management*. 1-57278-039-8 and 0-7844-0174-8. Water Environment Federation, Alexandria, VA.

Zielinski, J. (2000a). The benefits of better site design in residential subdivisions. In: *The Practice of Watershed Protection*, Schueler, T.R. and H.K. Holland, (editors). The Center for Watershed Protection, Ellicott City, MD. Pages 263 - 275.

Zielinski, J. (2000b). The benefits of better site design in commercial development. In: *The Practice of Watershed Protection*, Schueler, T.R. and H.K. Holland, (editors). The Center for Watershed Protection, Ellicott City, MD. Pages 277 - 286.

6 BMP Costs

Ari Selvakumar

CONTENTS

6.0 INTRODUCTION

Cost is an integral part of BMP design. The purpose of this chapter is to provide information on how to estimate the cost of structural BMPs. Costs associated with nonstructural BMPs are not included here as they are generally not as easily quantified as structural BMPs due to their indirect and highly variable implementation levels. The cost for constructing a structural BMP depends on many factors. They include the time of year, site conditions and topography, accessibility of equipment, economics of scale, and government regulations. Out of these factors, site conditions and topography are usually the most influential. Site preparation costs may be greatly reduced if existing conditions and vegetation are carefully integrated into the design of any BMP; e.g., a natural depression could be developed into some type of detention pond (Ferguson *et al.*, 1997).

It is normally less expensive from a unit cost standpoint to construct a larger project than a smaller one. This is because large portions of the costs are involved in mobilizing and demobilizing a project. Both overestimating and underestimating can occur by not considering the magnitude of the project (Ferguson *et al.*, 1997). However, a larger project usually involves significantly more land costs than does distributing BMPs into already set aside landscaping areas.

The cost of constructing any BMP is variable and can be substantial. Several documents have been published that address cost estimating for BMPs, but most of these report only construction costs (Young *et al.,* 1996; Sample *et al.*, 2003). In addition, costs are often documented as base costs and do not include land costs, which according to the U.S. EPA (1999) is the largest variable influencing overall BMP cost.

Design, operation, and maintenance (O&M) are other significant cost considerations discussed in only a few sources. Ferguson *et al.* (1997) published a fairly comprehensive BMP cost estimating guide as part of the Rouge River restoration project that includes construction and O&M costs for many BMPs. This guide could be used outside the Rouge River watershed provided adjustments are made for local labor rates and associated real estate costs. Young *et al.* (1996) compiled the results of past highway runoff research into a single volume users manual for highway practitioners which provide construction cost data for each BMP, but no cost data for design and maintenance. The Southeastern Wisconsin Regional Planning Commission (SWRPC) (1991) documents the most comprehensive analysis of construction and maintenance costs. The Water Environment Research Foundation (WERF) has summarized construction and maintenance costs for commonly used BMPs (WERF, 2003). The wide range of cost data reported in the literature indicates that much more information is needed in this area. Identifying the primary element that drives construction cost upward is a high priority and could help reduce the initial cost associated with BMPs.

In general, initial construction cost and maintenance cost were the factors considered in computing the cost effectiveness of BMPs. Recent studies have suggested that using construction and maintenance costs to compute cost effectiveness is insufficient. Landphair *et al.* of Texas Transportation Institute (2000) proposed that a real measure of cost effectiveness should include design, maintenance, and construction costs as well as pollutant removal performance of a selected BMP. Their research suggested that a cost index of this kind was only meaningful if special site considerations and land costs were ignored.

6.1 BMP Cost Estimating Methods

Four common methods of cost estimation are as follows (DOD, 1995):

- bottom-up method,
- analogy method,
- expert opinion method, and
- parametric method.

6.1.1 Bottom-Up Method

The bottom-up approach estimates costs on an item-by-item basis. Detailed methods typically rely on quantity take-offs and compiled sources of unit cost data for each item, taken from either a built-in database or other sources (e.g., cost-estimating references). This method is used when design information is available.

6.1.2 Analogy Method

This technique uses the cost of previously completed projects, as a comparison to the cost of the proposed project, to arrive at a final cost estimate. The actual costs from the completed project are extrapolated to estimate the cost of the proposed project (DOD, 1995). This method provides estimates by using the system or component level, which are obtained from using actual project costs from past experiences. The disadvantage of this method occurs when projects similar to the one in question do not exist, or the accuracy of the available data is questionable, or extrapolating the entire project cost does not accurately reflect evenly scaled costs.

6.1.3 Expert Opinion Method

Experts in this field can be consulted to provide a cost estimate for the project based upon their past experience and understanding of the proposed undertaking. The advantage of this is that the experts have the knowledge to account for the differences between past project experiences and the proposed project requirements, pointing out areas contributing to the cost estimate that may be overlooked. They can also factor in the impacts produced by new technologies or applications. The disadvantage of this technique is that the estimate is confined by the judgement and expertise of the consulted experts (DOD, 1995).

6.1.4 Parametric Method

The parametric approach relies on relationships between cost and design parameters. These relationships are usually statistically-based or model-based. Statistically-based approaches rely on "scaled-up" or "scaled-down" versions of projects where historical cost data is available. Model-based approaches utilize a generic design that is linked to a cost database and adjusted by the user for site-specific information. This method, also known as top-down estimating, is used when design information is not available (U.S. EPA, 2000).

6.2 BMP COST ANALYSES ELEMENTS

6.2.1 Total Costs

The total costs of BMPs include both capital (construction and land) and annual O&M costs. The first category of costs only occur in the year when the BMP is installed unless retrofits or up-sizing occur. The other costs may occur either yearly (e.g., operation cost) or some other frequency throughout the life of the BMP to maintain it.

6.2.1.1 Capital Costs

Capital costs are those expenditures that are required to construct a BMP. They are exclusive of costs required to operate or maintain the BMP throughout its lifetime. Capital costs consist primarily of expenditures initially incurred to build or install the BMP (e.g., land cost, construction of a wetland, and related site work).

Capital costs include all land, labor, equipment and material costs, excavation and grading, control structure, sediment control, landscaping, and appurtenances. Capital costs also include expenditures for professional/technical services that are necessary to support the construction of a BMP. The cost of constructing any BMP is variable and depends largely on site conditions and drainage area. For example, if a BMP is constructed in very rocky soils, the increased excavation costs may substantially increase the cost of construction. Land costs, which can be the largest variable influencing overall BMP cost, vary greatly from site to site.

Capital costs typically can be estimated using equations based on the size or volume of water to be treated (U.S. EPA, 1999). Stormwater practices have certain spatial requirements for surface area or volume necessary to treat a quantity of stormwater from a watershed and this variation is incorporated into the equations (U.S. EPA, 1999). These equations typically follow the single determinant equation as shown below. This equation is linear in the log transform.

$$C = aP^b$$

Where C is the estimated cost ($); P is the determinant variable (area or volume); and a and b are statistical coefficients determined from regression analysis. The exponent b represents the economics of scale factor. If $b < 1$, the unit cost decreases as size increases. When $b = 1$, the equation simplifies to a linear relationship and no economics-of-scale are present. If $b < 1$, then diseconomics of scale exist. Table 6-1 shows the unit costs of commonly used BMPs as reported in the literature. Unless indicated otherwise, all costs are expressed in terms of 2002 dollars based on the Engineering News Record (ENR) average annual cost index of 6538 (ENR, 2002). The cost equations show that the capital costs of most BMPs correlate well with basin volume. The same was noted by Koustas and Selvakumar (2003).

Traditionally, capital costs were calculated using the standard estimation guides such as the R.S.

Means Building Construction Cost Data Handbook used by many engineers. The Means book provides unit cost data for materials of construction, labor, equipment, installation, and excavation for cities across the U. S. However, linear equations are gaining popularity as they offer a way to replace a cost database with a single equation. These equations are based on data gathered from existing projects and assumptions that similar installations should cost about the same.

Table 6-1. Base Capital Costs (Without Land Costs) for Commonly Used BMPs

BMP Type	Base Capital Costs ($)	Reference
Detention Ponds/Dry Extended Detention Ponds	$C = 60{,}742V^{0.69}$; V in Mgal	Young *et al.*, 1996
	$C = 12.4V^{0.76}$; V in ft^3	Brown and Schueler, 1997
Wet Ponds/Retention Basins	$C = 67{,}368V^{0.75}$; V in Mgal	Young *et al.*, 1996
	$C = 24.5V^{0.71}$; V in ft^3	Brown and Schueler, 1997
Constructed Wetlands	$C = 30.6V^{0.71}$; V in ft^3	U.S. EPA, 2003
Infiltration Trenches/Filter Drains/Soakaways	$C = 173V^{0.63}$; V in ft^3	Young *et al.*, 1996
	$C = 5V$; V in ft^3	Brown and Schueler, 1997
Infiltration Basins	$C = 16.9V^{0.69}$; V in ft^3	Young *et al.*, 1996
Sand and Organic Filters	$C = KA$; A in acres; K ranges from 12,369 to 24,738	Young *et al.*, 1996
Vegetated Swales	$0.25 to $0.50/$ft^2$	WERF, 2003
Vegetated Buffer Strips	$0.30 to $0.70/$ft^2$	WERF, 2003
Porous Pavement	$2 to $3/$ft^2$	U.S. EPA, 2003
Bioretention	$3 to $4/$ft^2$	Coffman, 1999
	$C = 7.3V^{0.99}$; V in ft^3	U.S. EPA, 2003; Brown and Schueler, 1997
Water Quality Inlets (enhanced catch basins)	$8,000 to $24,000	Young *et al.*, 1996
	$2,000 to $3,000/basin for precast basins	U.S. EPA, 2003
	$400 to $10,000/basin for drop-in retrofits	U.S. EPA, 2003

Note: Costs in December 2002 dollars. Cost of land acquisition not included.
V = BMP Volume and A = BMP Area.

6.2.1.2 Design, Permitting, and Contingency Costs

Design and permitting costs include costs for site investigations, surveys, design and planning of a BMP (Novotny and Chesters, 1981). Contingency costs are simply any unexpected costs incurred during the development and construction of a BMP. These costs are usually expressed as a fraction of the base capital costs and have been considered as a uniform percentage for BMPs (U.S. EPA, 1999). Design, permitting, and contingency costs are reported in Table 6-2. These costs are generally only estimates, based on the experience of designers. Design, permitting and contingency costs can be a significant portion of the total capital cost of a BMP construction project particularly in areas where endangered species occur in contentious projects. In addition, cost of erosion and sediment control represents about 5 to 7% of the base capital costs (Brown and Schueler, 1997; U.S. EPA, 1999).

Table 6-2. Design, Permitting, and Contingency Costs

Additional Percentage (fraction of base capital costs)	Notes	Source
25 %	Includes design, permitting and contingency fees	Wiegand, *et al.* 1986; CWP, 1998; and U.S. EPA, 1999.
32 %	Includes design, permitting, contingency fees and erosion and sediment control costs	Brown and Schueler, 1997; U.S. EPA, 1999; and CWP, 1998.

6.2.1.3 Operation and Maintenance (O&M) Costs

O&M costs are those post-construction costs necessary to ensure or verify the continued effectiveness of a BMP. Like engineering design and construction, proper O&M is important in order to ensure that BMPs will function properly. These costs are estimated mostly on an annual basis.

Annual O&M costs include labor, materials, energy, and equipment required for proper operation and functionality of a BMP facility. These include costs for landscape maintenance, structural maintenance, infiltration maintenance, sediment removal, and basin debris and litter removal.

O&M costs are divided into two main categories: aesthetic and functional (U.S. EPA, 1999). Functional maintenance is important for performance and safety reasons, while aesthetic maintenance is important primarily for public acceptance of BMPs. Aesthetic appearance is more important for BMPs that are visible.

O&M costs are calculated on an annual basis throughout the life of a BMP facility. These costs are seldom available on a comprehensive basis and have been expressed as a fraction of the base capital costs. O&M costs, expressed as annual percentages of capital costs, as reported in literature, are presented in Table 6-3. O&M costs to maintain a stormwater BMP over a 20 to 25-year period are roughly equal to its initial construction cost (Weigand *et al.*, 1986). Few property owners and homeowner associations are fully aware of the need for and magnitude of stormwater maintenance costs, and most fail to regularly perform routine and non-routine maintenance tasks. It is likely that performance and longevity of many BMPs will decline without adequate maintenance. The information on the maintenance of BMPs is limited, even though the long term maintenance costs can be an important component of the life cycle cost of the facility.

Table 6-3. Annual Operation & Maintenance Costs

BMP Type	**Annual Maintenance Costs (% of Construction Cost without Land Cost)**	**Source**
Detention Ponds/Dry Extended Detention Ponds	<1%	Wiegand *et al.*, 1986; Schueler, 1987; SWRPC, 1991
Wet Ponds/Retention Basins	3 to 6%	Brown and Schueler, 1997; SWRPC, 1991
Constructed Wetlands	3 to 6%	Wiegand *et al.* 1986; Schueler, 1987; SWRPC, 1991
	2%	Livingston *et al.* 1997; Brown and Schueler, 1997
Infiltration Trench	5 to 20%	Schueler, 1987; SWRPC, 1991
Infiltration Basin	1 to 3%	Livingston *et al.* 1997; SWRPC, 1991
	5 to 10%	Wiegand *et al.* 1986; Schueler, 1987; SWRPC, 1991
Sand Filters	11 to 13%	Livingston *et al.* 1997; Brown and Schueler, 1997
Grassed Swales	5 to 7%	SWRPC, 1991
Vegetated Buffer Strips	$350/acre/year	SWRPC, 1991

6.2.1.4 Land Costs

Land cost is of critical importance because it constitutes a significant, if not major, component of total costs. The cost of land is site specific and extremely variable both regionally and by surrounding land use. For example, many suburban jurisdictions require open space allocations within the developed site, reducing the effective cost of land for BMPs to zero (Schueler, 1987). On the other hand, the cost of land may far outweigh construction and design costs in ultra urban settings. It should be noted that in some cases where a BMP can be incorporated into a planned or required landscaped area, there is no additional cost. For example, Portland, Oregon has a 15% landscaping requirement for multi-family, commercial, and light-industrial developments. Automatically, there is enough land for stormwater BMPs. The land requirement varies considerably depending on the type of BMP as shown in the Table 6-4 below.

Table 6-4. Relative Land Consumption of Stormwater BMPs

BMP Type	Land Consumption (% of Impervious Area of the Watershed)
Retention Basin	2 to 3%
Constructed Wetland	3 to 5%
Infiltration Trench	2 to 3%
Infiltration Basin	2 to 3%
Porous Pavement	0%
Sand Filters	0 to 3%
Bioretention	5%
Swales	10 to 20%
Filter Strips	100%

Source: U.S. EPA, 1999

6.2.1.5 Inflation and Regional Cost Adjustments

Any costs reported in the literature need to be adjusted for inflation and regional differences. To adjust for inflation, the ENR construction cost index history data can be used to adjust BMP cost data to current year dollars. The ENR index is published every month. The ENR index for the years 1990 to 2003 is shown in Table 6-5. To obtain current cost, the total capital BMP cost data is multiplied by current year index and divided by base year indexes. The total cost increase resulting from inflation can be calculated using the following equation:

$$\text{Capital Cost (adjusted)} = \frac{\text{Capital cost (base year) x ENR cost index for the current year}}{\text{ENR cost index for the base year}}$$

Table 6-5. ENR Annual Average Construction Cost Index History

Year	Construction Cost Index
1990	4732
1991	4835
1992	4985
1993	5210
1994	5408
1995	5471
1996	5620
1997	5826
1998	5920
1999	6059
2000	6221
2001	6343
2002	6538
2003	6695

Source: Engineering News Record (ENR, 2003)

Total capital costs for the construction of a BMP facility vary based on the region of the county or state in which the facility is located. Regional cost differences reflect differences in the cost of living, labor, and material costs for a particular location (U.S. EPA, 1999). Regional cost variations are perhaps the most difficult costs to estimate. Cost data can be adjusted for regional cost variation by using the ENR twenty city construction cost indexes for December 2002 (ENR, 2002). The twenty city construction cost index for December 2002 is shown in Table 6-6. The cost data adjusted for inflation is multiplied by the twenty city index factor (city index/national index for December 2002) of the city closest to the region of study to adjust for the regional variation.

Since the amount of regional rainfall may impact costs, further adjustment to cost is necessary. Based on a methodology presented by the American Public Works Association (APWA, 1992), a cost adjustment factor is assigned to each EPA rainfall region (shown in Figure 6-1). A regional cost adjustment factor based on regional rainfall zones is shown in Table 6-7. The cost adjusted for inflation and regional variation is reconciled for the rainfall bias by dividing the cost by the rainfall factor. For example, data for Florida (Region 3) is divided by 0.67 to adjust for the rainfall bias.

Table 6-6. Twenty City Construction Cost Index Regional Factors for December 2002

City	Index December 2002	Regional Factor
Atlanta	4189	0.64
Baltimore	4580	0.70
Birmingham	4686	0.71
Boston	7546	1.15
Chicago	7965	1.21
Cincinnati	6156	0.94
Cleveland	7067	1.08
Dallas	3895	0.59
Denver	4744	0.72
Detroit	7654	1.17
Kansas City	6782	1.03
Los Angeles	7403	1.13
Minneapolis	7621	1.16
New Orleans	3906	0.60
New York	10009	1.52
Philadelphia	8226	1.25
Pittsburgh	6419	0.98
St. Louis	7197	1.10
San Francisco	7644	1.16
Seattle	7562	1.15

Cost regional and inflation adjusted (December 2002) = Cost inflation adjusted (December 2002) x Regional Factor
Source: Engineering News Record (ENR, 2003); National Index for December 2002 is 6563

Table 6-7. Precipitation Regional Cost Adjustment Factors

Rainfall Zone	1	2	3	4	5	6	7	8	9
Adjustment Factor	1.12	0.90	0.67	0.92	0.67	1.24	1.04	1.04	0.76

Source: U.S. EPA, 1999

Note: This is very approximate as there are large differences in precipitation patterns.

6.2.2 Life Cycle Costs

Life cycle costs are costs incurred from "cradle to grave". Life cycle cost refers to the total project cost across the life span of a BMP, including design, construction, O&M, and closeout activities. It includes the initial capital costs and the present worth of annual O&M costs that are incurred over time, less the present worth of the salvage value at the end of the service life (Sample *et al.*, 2003). Life cycle cost analysis can be used to choose the most cost effective BMP from a series of alternatives so that the least long term cost is achieved.

A present worth analysis of future payment involves four basic steps:

1. define the period of analysis,
2. calculate the annual expenditures or cash outflow for each year of the project,
3. select a discount rate to use in the present worth calculation, and
4. calculate the present worth.

The present worth of a future payment is calculated using the following equation:

$$PW = \frac{x_t}{(1+i)^t}$$

where x_t is the payment in year t and i is the discount rate. The denominator, $(1 + i)^t$, is referred as a "discount factor." For a series of future payments, the total present worth from 1 to n years would be calculated as:

$$PW_{total} = \sum_{t=1}^{t=n} \frac{x_t}{(1+i)^t}$$

where, $\sum_{t=1}^{t=n} \frac{1}{(1+i)^t}$ is referred as the multi-year discount factor. When the annual cost, x_p, is constant over a period of years, the calculation of present worth can be simplified by using a multi-year discount factor.

Calculation Example - Wet Pond

In order to present how to calculate the life cycle cost of a BMP, a wet pond with a 1 acre-foot facility is selected. From Table 6-1, the construction cost for a wet pond can be calculated using the equation:

$$C = 24.5V^{0.71}$$

For a 1 acre-foot facility, the construction cost is equal to $45,700. From Table 6-2, the cost for design, permitting, and contingency is 25% of the construction cost, which is equal to $11,425.

For wet ponds, the annual cost of routine maintenance is typically estimated at about 3 to 6% of the construction cost (Table 6-3). Assuming 5%, the annual cost of maintenance becomes $2,285. Ponds are long-lived facilities (typically longer than 20 years). Assuming a period of analysis of 20 years and a discount rate of 7%, the multi-year discount factor is equal to 10.594. The present worth of maintenance for 20 years is calculated as 10.594 x $2,285 or $24,207.

Thus, for this example the life cycle cost equals the construction cost ($45,700), plus the design, permitting, and contingency cost ($11,425) and the present worth maintenance cost ($24,207) totaling $81,332. Land cost, which is extremely variable is not included.

6.3 ECONOMIC BENEFITS OF BMPS

BMPs can present several tangible economic benefits in spite of their construction, O&M costs. BMPs have the ability to mitigate potential downstream flooding associated with medium to larger storms. BMPs reduce flooding downstream from developments by reducing the peak flows and stretching out the flow over a long time. Avoiding floods eliminate property damages. Benefits of avoiding flood damages are relatively easy to estimate. Reducing peak flows can also potentially reduce capital and maintenance costs of drainage infrastructure.

BMPs can protect water resources downstream from urban development, which can prevent impact on human health through direct contact from swimming or through contamination of seafood. The economic benefits of avoiding human health problems include swimming and recreational benefits as well as saved medical costs. One study in Saginaw, Michigan estimated that the swimming and beach recreation benefits associated with a CSO retention project exceeded seven million dollars (U.S. EPA, 1998). BMPs also reduce the cost of water quality improvement.

BMPs are beneficial to developers as they create a "water" feature effect and people have a strong emotional attachment to water, arising from its aesthetic qualities– tranquility, coolness, and beauty. BMPs within developments can be used as marketing tools to set the tone for the entire project.

BMPs that are visually aesthetic and safe for children can lead to increased property values. As such, beautification of land areas adjacent to BMPs should be considered an integral part of planning by developers. It was found that developers could charge per lot a premium of up to $10,000 for homes situated next to well-designed stormwater ponds and wetlands (U.S. EPA, 1995). A study conducted by the National Association of Home Builders indicates that the proximity to water raises the value of a home by up to 28% (U.S. EPA, 1995). If the BMPs are developed to allow passive recreation (e.g., a walking path around a lake or pond), the recreational area and the BMP can become the feature attraction when advertising the property. BMPs with standing water often appear to be natural systems. Many developers have capitalized on urban runoff regulations by designing aesthetic wet ponds and stormwater wetlands and marketing them as if they were natural lakes, ponds, or wetlands. In addition, U.S. EPA found that office parks and apartments next to a well-designed BMP could be leased or rented at a considerable premium and often at a much faster rate. BMPs may require more than periodic maintenance to preserve environmental and monetary benefits of "waterfront" lots. However, the benefits of higher resale value and quality of life typically outweigh the combined costs of the initial lot premium and annual maintenance fees charged by homeowners' associations (U.S. EPA, 1995).

Stormwater BMPs such as grassed swales and bioretention areas are actually less expensive to construct than enclosed storm drain systems and they also provide better environmental results (NJDEP, 2000). Also, the more natural drainage systems eliminate the need for costly manholes, pipes, trenches, and catch basins, while removing pollutants at the same time.

Stormwater BMPs must be maintained, and the cost burden falls on landowners or local government. Over a 20 to 25 year period, the full cost to maintain a stormwater BMP is roughly equal to its initial construction cost (Wiegand *et al.*, 1986). Poorly maintained BMPs may be unsightly due to excess algal growth or public littering. Few property owners and home owners' associations are fully aware of the magnitude of stormwater BMP maintenance costs, and fail to perform maintenance tasks.

Abatement or prevention of pollution from storm-generated flow is one of the most challenging areas in the environmental engineering field. The facts of life - from an engineering standpoint are difficult to face in terms of design and cost. Operational problems can be just as foreboding.

The full impacts of "marginal" pollution, particularly that caused by uncontrolled overflows, must be recognized now and planning initiated to improve drainage system efficiencies and bring all wastewater flows under control. Municipal programs with this objective cannot begin too soon because corrective action is time-consuming. Efforts devoted to improved sewerage and drainage systems will pay significant dividends in complete control of metropolitan wastewater problems and pollution abatement. Research and development are making available important answers on the most efficient and least costly methods needed to restore and maintain water resources for maximum usefulness to man.

It is clear that abatement requirements for storm-flow pollution are here. Already, federal and local governments have promulgated wet-weather flow treatment and control standards. Newly developed and developing watershed districts can take a crucial opportunity and assess what has transpired and determine their own best water management strategy.

To exemplify this, one can consider the water pollution control efforts in the U. S. Historically, this nation has always approached water pollution control in a series and segmented manner with respect to time and pollutant sources, respectively. The result is that the problem is still being fought after more than 60 years of effort and billions of dollars of expenditures. Initially, they abated sanitary sewage; first with primary treatment, and later only after a long time, with secondary treatment. Somewhere in between attempts to control sanitary sewage, industrial wastewater control became a requirement; however, pretreated industrial wastewaters are still released during an overflow event. Only recently were we forced to control combined sewer overflow, and now we are faced with requirements to abate separate-stormwater pollution. The aforementioned historical approach to water pollution control has taken a very long time, and only after trial and error of each individualized and fragmented approach was it learned that receiving water pollution problems remain. If instead, an entire watershed or multi-drainage area analysis was conducted earlier, a determination could have been made of the overall pollution problem in the receiving water bodies, the pollution sources (or culprits) contributing to the problem, and an optimized, integrated, area-wide program to correct the problem.

After the macro- (or large-scale watershed) analysis is conducted, an optimized determination of what sources to be abated (or where to spend the monies) will be made. Then, with the resulting information, a micro- (or drainage area/pollutant source and control) analysis can be performed.

There is one other important consideration that must be made, i.e., the reuse and reclamation of stormwater for such beneficial purposes as aesthetic and recreational ponds, groundwater recharge, irrigation, fire protection, and industrial water supply.

An optimal approach to integrated stormwater management is a total watershed or basin-wide analyses including a macro- or large-basin-scale evaluation interfaced with a descretized micro- or small-catchment-scale evaluation involving the integration of: (1) all catchments or drainage areas, tributaries, surrounding water bodies, and groundwater; (2) all pollutant source areas, land uses, and flows; and (3) added storm-flow sludge and residual solids handling and disposal. Flood and erosion control along with reuse and reclamation technology must also be integrated with pollution control, so that the retention and drainage facilities required for flood and erosion control can be simultaneously designed or retrofitted for pollution control and stormwater reclamation.

In conclusion, knowledge of interconnecting basinwide waters and pollutant loads affecting the receiving water body and the subsurface and groundwater will result in knowing how to get the optimum water resource and pollution abatement and a much more expedient and cost-effective water management program.

REFERENCES

American Public Works Association (APWA). 1992. A Study of Nationwide Costs to Implement Municipal Storm Water Best Management Practices. Southern California Chapter. Water Resource Committee.

Brown, W. and T. Schueler. 1997. The Economics of Storm Water BMPs in the Mid-Atlantic Region. Center for Watershed Protection. Elliot City, MD, 41 pp.

Center for Watershed Protection (CWP). 1998. Cost and Benefit of Stormwater BMPs. Final Report 9/14/98. EPA Contract 68-C6-0001 WA2-15, Task 6 Parson Engineering Science.

Coffman, L., M. Clar, and N. Weinstein. 1998. Overview: Low Impact Development for Stormwater Management. Unpublished.

ENR. 2002. Construction Economics. pp. 23. December 9.

ENR. 2003. Construction Cost Index History. http://www.enr.com/features/conEco/costIndexes/constIndexHist.asp

Ferguson, T., R. Gignac, M. Stoffan, A. Ibrahim, and J. Aldrich. 1997. Rouge River National Wet Weather Demonstration Project: Cost Estimating Guidelines, Best Management Practices and Engineered Controls. Wayne County, MI.

Koustas, R.N. and A. Selvakumar. 2003. BMP Cost Analysis for Source Water Protection. American Water Resources Association (AWRA) 2003 International Conference in New York in June.

Landphair, H.C., J.A. McFalls, and D. Thompson. 2000. Design Methods, Selection, and Cost-Effectiveness of Stormwater Quality Structures. Report No. FHWA/TX-01/1837-1. Texas Transportation Institute, The Texas A&M University System, College Station, Texas 77843.

NJDEP. 2000. Revised Manual for New Jersey: Best Management Practices for Control of Nonpoint Source Pollution from Stormwater. http://www.state.nj.us/dep/watershedmgt/bmpmanual.html

Novotny, V. and G. Chesters. 1981. Handbook of Nonpoint Pollution Sources and Management. Van Nostrand and Reinhold Environmental Engineering Series, New York, NY.

Sample, D.J., J.P. Heaney, L.T. Wright, C.Y. Fan, F.H. Lai, and R. Field. 2003. Cost of Best Management Practices and Associated Land for Urban Stormwater Control. Journal of Water Resources Planning and Management, Vol. 129, No. 1, pp. 59-68.

Schueler, T.R. 1987. Controlling Urban Runoff: A Practical Manual for Planning and Designing Urban BMPs. Metropolitan Washington Council of Governments, Washington, DC.

Southeastern Wisconsin Regional Planning Commission (SWRPC). 1991. "Costs of Urban Nonpoint Source Water Pollution Control Measures". Technical Report No. 31 Waukesha, WI, 107 pp.

U.S. Department of Defense (DOD). 1995. Parametric Cost Estimating Handbook. Joint Government/Industry Initiative.

U.S. EPA. 1990. "National Pollutant Discharge Elimination System Permit Application Regulations for Storm Water Discharges." (Phase I Final Rule). Federal Register 55 FR 47990.

U.S. EPA. 1995. Economic Benefits of Runoff Controls. EPA-841-S-95-002. Office of Wetlands, Oceans, and Watersheds, Washington, DC.

U.S. EPA. 1998. National Pollutant Discharge Elimination System–Proposed Regulations for Revision of the Water Pollution Control Program Addressing Storm Water Discharges. (Phase II Proposed Rule). *Federal Register.* 63FR1536, January 8.

U.S. EPA. 1999. Preliminary Data Summary of Urban Storm Water Best Management Practices. EPA-821-R-99-012. August. Office of Water, Washington, DC.

U.S. EPA. 2000. A Guide to Developing and Documenting Cost Estimates During the Feasibility Study. EPA/540/R-00/002. OSWER 9355.0-75. July.

U.S. EPA. 2003. *National menu of best management practices for storm water Phase II.* Office of Water, Washington, DC. http://cfpub.epa.gov/npdes/stormwater/menuofbmps/menu.cfm.

WERF. 2003. Post-Project Monitoring of BMPs/SUDS to Determine Performance and Whole-Life Costs. Project Number: 01-CTS-21T.

Wiegand, C., T. Schueler, W. Chittenden, and D. Jellick. 1986. Cost of Urban Runoff Quality Controls. pp. 366-380. In: Urban Runoff Quality. Engineering Foundation Conference. ASCE, Henniker, NH. June 23-27.

Young, G.K., S. Stein, P. Cole, T. Kammer, F. Graziano, and F. Bank. 1996. Evaluation and Management of Highway Runoff. Water Quality Technical Report. Department of Environmental Programs, Metropolitan Washington Council of Governments, Washington, DC.

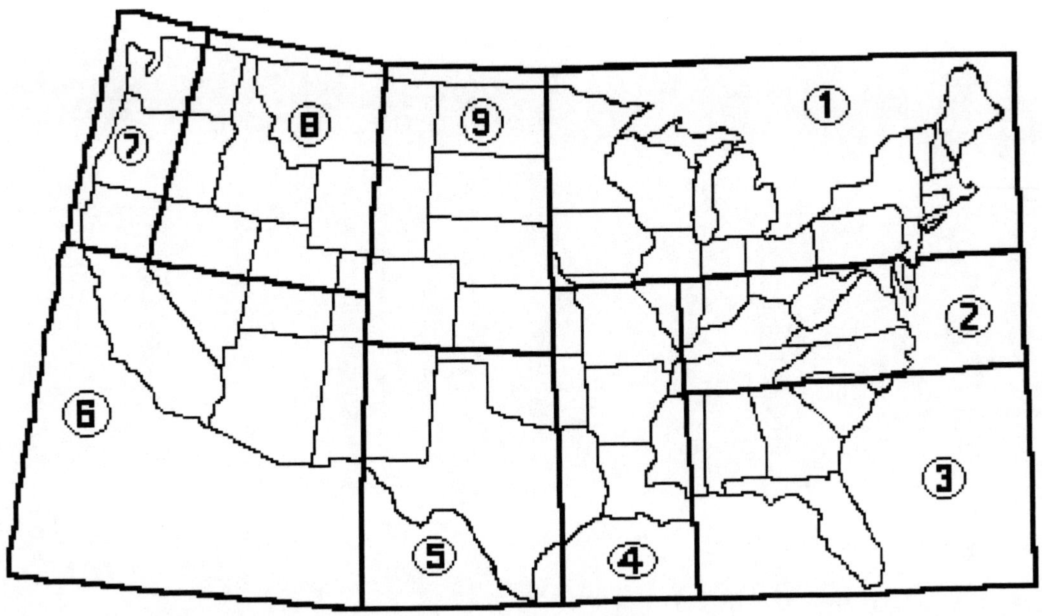

Figure 6-1. Regional Rainfall Zones for the United States

Rainfall Zones can be found in the NPDES Phase I regulations, 40 CFR Part 122, Appendix E (U.S. EPA, 1990)

Index

Contributors

Richard Field, who has worked in the environmental engineering field for 42 years, received a bachelor of civil engineering degree from the City College of New York (CCNY) in 1962 and a master's degree in civil engineering (sanitary engineering option) from New York University (NYU) in 1963, graduating first in his class. He has since taken many postgraduate courses related to environmental engineering, construction technology, advanced mathematics, computer technology etc. He is a registered professional engineer (P.E.) in the states of New York and New Jersey, a member of Chi-Epsilon National Civil Engineering Honor Fraternity, a member of the American Society of Civil Engineers (ASCE) and an executive committee member of its Urban Water Resources Research Council. He has been a member of the following committees: The Water Environment Federation (WEF) Water Environment Research Foundation Wet Weather Advisory Panel and Research Committee covering urban wet weather flows, the Environment Canada (the Canadian federal environmental agency) Steering Committee on CSO High-Rate Treatment and the U.S. Environmental Protection Agency (U.S. EPA) Sanitary Sewer Overflow (SSO) Advisory Committee and Urban Wet Weather Flow Subcommittee.

Since May 1970, Mr. Field has been in charge of the U.S. EPA National Storm and CSO (combined sewer overflow) Technology Research and Development Program located at the National Risk Management Research Laboratory in Edison, New Jersey. He has received numerous outstanding-achievement awards and citations for on-the-job performance and technological contribution including two U.S. EPA bronze medals, the ASCE State-of-the-Art of Civil Engineering Award, two New York Water Pollution Control Association awards for excellence in technological advancement, three U.S. EPA Scientific and Technological Achievement awards including a first-level award, and a first place U.S. EPA National Award in the CSO category. He has authored and coauthored, presented, or published a combination of more than 800 peer-reviewed articles, conference proceedings, papers, other papers, U.S.EPA reports, books and book chapters, some of which are internationally recognized publications in his field. He has been invited to lecture and present seminars throughout the world and has presented more than 300 times. Mr. Field is a U.S. EPA expert and an internationally recognized expert in urban wet weather discharge impacts and control technology including the areas of CSO, SSO, and infiltration/inflow (I/I), urban stormwater, diffuse or nonpoint sources, and watershed management. He is listed in Who's Who in Engineering, Who's Who in Technology Today, Who's Who in Science and Engineering, Who's Who in Finance and Industry, Who's Who in the East, International Who's Who of Professionals and Who's Who in the World.

Anthony N. Tafuri received a bachelor of science degree in engineering from Hofstra University in 1964, and master's degrees in civil engineering from Columbia University in 1966 and (sanitary engineering option) from New York University (NYU) in 1971. He is a registered professional engineer in the states of New York and New Jersey and a member of Chi-Epsilon, the National Civil Engineering Honor Fraternity.

Mr. Tafuri has worked in the environmental field for 38 years. He has been involved in several program areas since his research career started at the U.S. EPA in 1971. From 1971 to 1980 he was an environmental engineer in the U.S. EPA's Storm and Combined Sewer Technology Program and worked on the development and demonstration of technologies to abate pollution from storm and combined sewer discharges and non-sewered urban runoff. From 1980 to 1995, he researched technologies to clean up and abate pollution from uncontrolled hazardous waste sites. For the last 10 years of this period he managed the U.S. EPA's Underground Storage Tank Program and was instrumental in providing technical support for the development and implementation of regulations for the prevention, detection and cleanup of contamination from leaking tanks and pipelines containing petroleum products. Since 1995, Mr. Tafuri has been working in the U.S. EPA's Urban Watershed Management Program and has been chief of the Urban Watershed Management Branch since September 2004, addressing the problems associated with the collection, control, and treatment of wet-weather flow within the urban watershed. He is also involved in research activities associated with aging and obsolete water and wastewater infrastructure and in the U.S. EPA's Environmental Technology Verification (ETV) Program.

Mr. Tafuri has received numerous awards and citations for on-the-job performance and technological achievements, including four U.S. EPA Bronze Medals for outstanding contributions to the U.S. EPA's storm water runoff pollution abatement program, underground storage tank program, and hazardous waste program. He is the recipient of Vice President Gore's Hammer Award for his teamwork in developing the environmental technology verification program. He is also the recipient of a U.S. EPA Scientific and Technological Achievement Award for his research publication on locating leaks in underground pipelines with acoustic technology and was nominated for U.S. EPA's Science Achievement Award for exceptional leadership and administration of U.S. EPA's Leaking Underground Storage Tank Program. During his research career, Mr. Tafuri has authored, coauthored, presented and published a combination of more than 250 peer-reviewed technical documents, including books and book chapters on wet-weather flow technology, drinking-water distribution systems, hazardous waste control, and corrective actions for leaking underground storage tank sites.

Swarna Muthukrishnan received her Ph.D. in Environmental Sciences from Jawaharlal Nehru University, New Delhi, India, in 2001, and M.S. in Environmental Chemistry from University of Madras, India, in 1995, graduating first in the University during the M.S. Program. She also holds a B.S. in Chemistry from the University of Madras, India, awarded in 1993. She is presently working as a Post Doctoral Fellow under the ORISE Program (Oak Ridge Institute for Science Education) at U.S. EPA's Office of Research and Development, Urban Watershed Management Branch, Edison, New Jersey, and is engaged in research efforts on urban stormwater management using structural best management practices (BMPs). A member of the American Chemical Society, Dr. Muthukrishnan's professional interests include environmental analytical chemistry and biogeochemical processes of aquatic, estuarine, wetland, and terrestrial systems. Her specific research interests focus on the chemical speciation of heavy metals in natural waters, soils, and sediments, as well as investigating the use and evaluation of urban stormwater BMPs to protect and improve the quality of receiving waters.

Bethany AnnMadge Acquisto, Ph.D., is currently a federal post-doctoral fellow for the U. S. Environmental Protection Agency (U.S. EPA), Office of Research and Development (ORD), National Risk Management Research Laboratory (NRMRL), Technology Transfer and Support Division (TTSD), Technology Transfer Branch (TTB) in Cincinnati, Ohio. She is investigating issues pertaining to the land application of biosolids and their possible pathological impacts. While working on this book she was an Oak Ridge Institute for Science Education post-doctoral fellow assigned to the U.S. EPA/ORD/ RMRL/Water Resources and Water Supply's Division's Urban Watershed Management Branch in Edison, NJ, where her main focus was the study of particulate-bound nutrients in urban runoff. Prior to her post-doctoral positions, Bethany was a research assistant and doctoral candidate at the University of Buffalo, Buffalo, NY. She received her Ph.D. in February 2003 in Civil/Environmental Engineering. Her dissertation research was in ultraviolet disinfection and more specifically the interference of bacteria/particle associations.

Ariamalar Selvakumar has more than 15 years of experience working in the environmental engineering field. She is currently an environmental engineer with U.S. EPA Urban Watershed Management Branch in Edison, New Jersey. She conducts research with urban stormwater runoff. Prior to joining the U.S. EPA in 1999, she worked for Foster Wheeler Corporation for about 9 years. She holds a Ph.D. in Civil and Environmental Engineering from New Jersey Institute of Technology (1989) and an M.Eng. in Environmental Engineering from Asian Institute of Technology, Bangkok, Thailand (1984). She also holds a B.Eng. (Hons.) in Civil Engineering from University of Peradeniya, Sri Lanka (1980). She is a registered professional engineer in the State of New Jersey. She has authored, co-authored, or presented a combination of more than 40 peer-reviewed journal articles, conference proceedings, posters, and U.S. EPA reports. She has received awards and citations for on-the-job performance and technical achievements including an honorable mention in U.S. EPA Scientific and Technological Achievement Awards in 2003. She is listed in Who's Who in America 2005.

Daniel Sullivan received a bachelor of science in civil engineering from the Polytechnic Institute of Brooklyn in 1968 and a master's degree in science in environmental sciences and engineering from the University of North Carolina in 1970. He has worked in the environmental engineering field for 32 years, is a registered professional engineer in the states of New York and New Jersey, a professional planner in New Jersey and a member of Chi Epsilon, the National Civil Engineering Honor Fraternity. From 1995 till his retirement in January 2004, Mr. Sullivan was Chief of the Urban Watershed Management Branch which conducts the U.S. EPA national wet-weather flow and watershed management research program. He began his U.S. EPA career in 1972 and has authored and coauthored, presented, or published a combination of more that 100 peer-reviewed articles, conference proceedings, papers, U.S. EPA reports, books, and book chapters on topics of hazardous waste control and wet-weather flow. He received two U.S. EPA bronze medals for work in the U.S. EPA wet-weather flow research and environmental technology verification programs.

The Use of Best Management Practices (BMPs) in Urban Watersheds

DEStech Publications, Inc.
439 North Duke Street
Lancaster, Pennsylvania 17602 U.S.A.

Printed in the United States of America
10 9 8 7 6 5 4 3 2

Main entry under title:
The Use of Best Management Practices (BMPs) in Urban Watersheds

A DEStech Publications book
Bibliography: p.
Includes index p. 261

ISBN: 1-932078-46-0
ISBN13: 978-1-932078-46-6

The Use of Best Management Practices (BMPs) in Urban Watersheds

Edited by:
Richard Field
Anthony N. Tafuri
Swarna Muthukrishnan
Bethany AnnMadge Acquisto
Ariamalar Selvakumar

DEStech Publications, Inc.